CW01456097

U.S.
BATTLESHIPS

A COMPLETE ILLUSTRATED HISTORY
[FROM BB-1 to BB-64]

Edited By
Marc Schulman

U.S. BATTLESHIPS
A COMPLETE ILLUSTRATED HISTORY
[FROM BB-1 to BB-64]

Edited by Marc Schulman

Published by MultiEducator PRESS, New York
177A East Main Street • New Rochelle, NY 10801
Email: info@historycentral.com

© 2024 HistoryCentral Press
ISBN # 978-1-885881-82-3

© 2024 All rights reserved by the author Marc Schulman

The right of Marc Schulman to be identified as editor of this work has been asserted (with exceptions in the case of the reading excerpts that are reprinted here with permission), in accordance with the US 1976 Copyright 2007 Act. No part of this book may be reproduced or utilized in any form or by any means, electronic or mechanical, or by any information storage and retrieval system without the prior permission of the publisher. The only exception to this prohibition is "fair use" as defined by U.S. copyright law.

Table of Contents

Preface

This book is a comprehensive exploration of the United States' journey through the age of the battleship, from the first of its kind, BB-1, to the last, BB-64. It offers an in-depth look at the pivotal role these leviathans of the sea played in shaping maritime strategy and national security. The histories presented are meticulously drawn from the US Navy's authoritative "History of US Fighting Ships," with only mild edits to enhance readability and ensure a coherent narrative thread that spans over a century of naval innovation and bravery.

The images within these pages bring the past into vivid color, all sourced from the esteemed collections of the National Archives and the Library of Congress. These photographs, originally in black and white, have been carefully colorized to breathe life into the scenes, allowing readers to appreciate the grandeur of these vessels in their full glory.

A significant portion of this book delves into the evolution of the different classes of American battleships. It charts the technological advancements and strategic doctrines that influenced their design and deployment, underscoring how each class was a response to the growing demands of naval warfare and a testament to American ingenuity and foresight. From the coastal defense ships of the late 19th century to the formidable Iowa-class battleships, this section not only highlights the ships' technical specifications and armaments but also their central role in the United States' naval strategy up to World War II.

During the Second World War, the role of battleships underwent a pivotal shift. While they participated in some of the conflict's defining naval battles, their role increasingly leaned towards shore bombardment, providing invaluable support to ground forces during key amphibious assaults. This transition marked a change in naval warfare, reflecting the growing importance of air power and the emergence of new technologies that would define the future of naval engagements.

The decommissioning of the last battleship in 1994 marked the end of an era. However, the legacy of these vessels continues to captivate and inspire. This book is not only a tribute to the technological marvels that these ships were but also a homage to the men who served on them, whose bravery and dedication played a crucial role in securing the seas for the United States.

Through this narrative, we aim to provide readers with a deeper understanding of the significance of these battleships, not just as instruments of war, but as symbols of American resilience and innovation.

BB-1 • USS INDIANA

Indiana, the 19th State, was admitted to the Union 11 December 1816, and was named because it had been the home of Indians.

(BB-1: dp. 10,288 n.; l. 350'11"; b. 69'3"; dr. 24'; s. 15 k.; cpl. 473; a. 4 13", 8 8", 4 6", 20 6-pdr., 6 1-pdr.)

The first Indiana (BB-1) was laid down on 7 May 1891 by William Cramp & Sons, Philadelphia; launched on 28 February 1893; sponsored by Miss Jessie Miller, daughter of the Attorney General of the United States, and commissioned on 20 November 1895 with Captain Robley D. Evans in command. Following fitting out at Philadelphia Navy Yard, Indiana trained off the coast of New England. This duty continued until the outbreak of the Spanish-American War in 1898, when Indiana formed part of Admiral Sampson's squadron. The 10 ships sailed south to intercept Cervera's Spanish squadron, known to be en route to the Caribbean. Indiana took part in the bombardment of San Juan on 12 May 1898, and returned to Key West with the squadron to guard Havana on 18 May. After it was discovered that Cervera was at Santiago, Sampson joined Schley there on 1 June and took up the blockade.

In late June, Army units arrived and were landed for an assault on Santiago. Cervera saw that his situation was desperate and began his gallant dash out of Santiago on 3 July 1898, hoping to outrun the American blockaders. Indiana did not join in the initial chase because of her extreme eastern position on the blockade but was near the harbor entrance when destroyers Pluton and Furor emerged. In a short time, both ships were destroyed by Indiana's guns and those of the other ships. Meanwhile, the remaining Spanish vessels were sunk or run ashore, in one of the two major naval engagements of the war.

Indiana returned to her previous pattern of training exercises and fleet maneuvers after the war and made practice cruises for midshipmen of the Naval Academy before decommissioning on 29 December 1903.

The battleship recommissioned at New York Navy Yard on 9 January 1906. During this phase of her career, Indiana served with the Naval Academy Practice Squadron, sailing to Northern Europe and the Mediterranean. At Queenstown, Ireland, she fired a 21-gun salute on 22 June 1911 in honor of

the coronation of King George V. This important work in training the Navy's future leaders ended in 1914, and she decommissioned at Philadelphia on 23 May 1914.

Indiana recommissioned a second time on 24 May 1917 and served through World War I as a training ship for gun crews off Tompkinsville, N.Y., and in the York River, Va. She decommissioned at Philadelphia on 31 January 1919. The name Indiana was canceled on 29 March 1919, and she was reclassified as Coast Battleship Number 1 so that the name could be assigned to a newly authorized battleship. She was used as a target in an important series of tests designed to determine the effectiveness of aerial bombs and was sunk in November 1920. Her hulk was sold for scrap on 19 March 1924.

BB-2 • USS MASSACHUSETTS

(BB-2: displacement 10,288; length 350'11"; beam 69'3"; draft 24'; speed 16.21 knots; complement 586; armament 4 13-inch guns, 8 8-inch guns, 4 6-inch guns, 2 3-inch guns, 20 6-pounders, 6 1-pounders, 6 18-inch torpedo tubes; class Indiana)

The fourth ship named Massachusetts (BB-2) was laid down by William Cramp & Sons in Philadelphia, Pennsylvania, on June 25, 1891. It was launched on June 10, 1893, sponsored by Miss Leila Herbert, daughter of the Secretary of the Navy Hilary Herbert, and commissioned on June 10, 1896, with Captain Frederick Rodgers in command.

Setting sail for shakedown on August 4, 1896, Massachusetts conducted trials and maneuvers off the middle Atlantic coast until November 30, when she entered the New York Navy Yard for an overhaul. After a brief voyage to Charleston, South Carolina, from February 12 to 20, 1897, the battleship departed New York on May 26 for Boston, arriving two days later. There, a celebration was held in her honor, including the presentation of the Massachusetts Coat of Arms on June 16, and a gift of a Victory statue the next day. She left Boston on June 15 to cruise to St. Johns, Newfoundland, arriving on June 23. Departing on June 28, the warship operated off the Atlantic coast for the next ten months, participating in training maneuvers with the North Atlantic Squadron off Florida and visiting major East Coast ports. On March 27, 1898, she was ordered to

Hampton Roads, Virginia, to join the "Flying Squadron" for the blockade of Cuba.

Massachusetts departed Norfolk on May 13 for Cienfuegos, Cuba, and began blockade duties on May 22. On the afternoon of May 31, she, along with the battleship Iowa (BB-4) and cruiser New Orleans, bombarded forts at the entrance to Santiago de Cuba, and exchanged fire with the Spanish cruiser Cristobal Colon, forcing the enemy ship to retreat into Santiago's inner harbor. The battleship remained on patrol off Santiago, intermittently bombarding Spanish fortifications until July 3, when she left to coal at Guantanamo Bay. Missing the Battle of Santiago, she returned to her station on July 4, arriving in time to assist the battleship Texas in forcing the cruiser Reina Mercedes to beach and surrender at midnight on July 6. After supporting the American occupation of Puerto Rico from July 21 to August 1, Massachusetts returned to New York, arriving on August 20.

Over the next seven years, Massachusetts cruised the Atlantic coast and the eastern Caribbean as a member of the North Atlantic Squadron. From May 27 to August 30, 1904, she served as a training ship for Naval Academy midshipmen off New England, then entered the New York Yard for an overhaul. Departing New York on January 13, 1905, she steamed for the Caribbean for training maneuvers and returned north to cruise off New England in May. Entering New York on November 12, 1905, she underwent an inactivation overhaul and decommissioned on January 8, 1906.

The battleship was placed in reduced commission on May 2, 1910, to serve as a summer practice ship for Naval Academy midshipmen. Over the next four years, she made three midshipman cruises—twice to Western Europe—before joining the Atlantic Reserve Fleet in September 1912. After a brief voyage to New York from October 5 to 16 for the Presidential Fleet Review, the warship returned to Philadelphia, where she remained until decommissioning on May 23, 1914.

Massachusetts was recommissioned on June 9, 1917, in Philadelphia. Sailing on October 9, she arrived at the Naval Training Station in Newport, Rhode Island, the same day, where she embarked Naval Reserve gun crews for gunnery training in Block Island Sound. Continuing this duty until May 27, 1918, the old battleship then underwent repairs at the Philadelphia

Navy Yard. Assigned to battle practice, "A" Division, Battleship Force 1, Atlantic Fleet, on June 9, 1918, she steamed to Yorktown, Virginia, the same day. For the remainder of World War I, she served as a heavy gun target practice ship in Chesapeake Bay and local Atlantic waters. Massachusetts returned to Philadelphia on February 16, 1919. Redesignated Coast Battleship No. 2 on March 29, she was decommissioned for the final time on March 31. Struck from the Navy list on November 22, 1920, she was loaned to the War Department as a target ship. After being scuttled off Pensacola Bar, Florida, on January 6, 1921, the hulk was bombarded by batteries from Fort Pickens for four years before being returned to the Navy on February 20, 1925. Though offered for sale for scrap, no acceptable bids were received, and finally, on November 15, 1956, the ship was declared the property of the state of Florida

BB-3 • USS OREGON

(Battleship No. 3: dp. 11,688; l. 351'2"; b. 69'3"; dr. 24'0"; s. 16 k.; cpl. 473; a. 4 13", 8 8", 4 6", 20 6-pdr., 6 1-pdr., 6 18" tt. (surf.); cl. Indiana)

Oregon (Battleship No. 3) was laid down on November 19, 1891, by Union Iron Works in San Francisco, California. She was launched on October 26, 1893, sponsored by Miss Daisy Ainsworth, and commissioned on July 15, 1896, with Captain Henry L. Howison in command.

After commissioning, Oregon was fitted out for duty on the Pacific Station, where she served for a short time. Leaving dry dock on February 16, 1898, she received news that Maine had exploded in Havana harbor the previous day. As tensions with Spain grew, Oregon arrived in San Francisco on March 9 and loaded ammunition. Three days later, she was ordered on a historic voyage around South America to the East Coast for action in the impending war with Spain.

Oregon departed San Francisco on March 19 for Callao, Peru, the first coaling stop on her journey. Arriving in Callao on April 4 and departing several days later, her commanding officer, Captain Charles E. Clark, elected not to stop at Valparaiso, Chile, for coal but to continue through the Straits of Magellan. On April 16, Oregon entered the Straits and encountered a severe gale, obscuring the rocky coastline. She anchored on a rocky shelf near Cape Forward and safely weathered the night. By dawn on April 17, the gale moderated, and Oregon proceeded to Punta Arenas, where she was joined by the gunboat Marietta, also sailing to the East Coast.

Both ships coaled and departed on April 21 for Rio de Janeiro, keeping their guns manned due to a rumored Spanish torpedo boat in the area. Delayed by head seas and winds, they didn't reach Rio until April 30. There, Oregon received news of the declaration of war against Spain and departed on May 4. With a brief stop in Bahia, Brazil, she arrived in Barbados for coal on May 18, and anchored off Jupiter Inlet, Florida, on May 24, reporting ready for battle. Oregon had sailed over 14,000 miles since leaving San Francisco 66 days earlier. The voyage demonstrated the military capabilities of a heavy battleship in all conditions and underscored the need for the Panama Canal.

On May 26, Oregon proceeded to the Navy Base at Key West, joined Admiral Sampson's fleet two days later, and arrived off Santiago, Cuba, on June 1 to shell military installations and help destroy Admiral Cervera's fleet on July 3. Oregon then went to the New York Navy Yard for a refit and, in October, sailed for the Asiatic station.

Arriving in Manila on March 18, 1899, Oregon remained in the area until the following February. She cooperated with the Army during the Philippine Insurrection, performing blockade duty in Manila Bay and off Lingayen Gulf, serving as a station ship, and aiding in the capture of Vigan.

Departing Cavite on February 13, 1900, Oregon cruised in Japanese waters until May, then went to Hong Kong. Ordered to proceed to Taku for the Boxer Rebellion, she departed on June 23 and grounded on an uncharted rock in the Straits of Pechili on June 28. Damaged and taking on water, Oregon was towed to Hope Sound for temporary repair and then to Kure, Japan, for final repairs.

Resuming her journey to China on August 29, 1900, Oregon cruised off the Yangtze River and served as a station ship at Woosung. Departing on May 5, 1901, she sailed via Yokohama and Honolulu, arriving in San Francisco on June 12, and entered Puget Sound Navy Yard on July 6 for an overhaul.

Oregon remained in the Puget Sound area for over a year and returned to Asiatic waters on March 18, 1903. She visited various Chinese, Japanese, and Philippine ports before returning to the West Coast in February 1906 and decommissioning at the Puget Sound Navy Yard on April 27.

Oregon recommissioned on August 29, 1911, but remained in reserve until October when she sailed to San Diego. The following years were relatively inactive, as she operated out of West Coast ports. On April 9, 1913, she was placed in ordinary at Bremerton, Washington, and on September 16, 1914, entered a reserve status, although still commissioned. Fully commissioned again on January 2, 1915, Oregon sailed to San Francisco for the Panama-Pacific International Exposition. From February 11, 1916, to April 7, 1917, she was in reserve at San Francisco. Returning to full commission on April 7, 1917, Oregon first operated on the West Coast, then escorted transports for the Siberian Expedition. After World War I, she decommissioned at Bremerton on June 12, 1919. From August 21 to October 4, she recommissioned briefly and served as the reviewing ship for President Woodrow Wilson during the arrival of the Pacific Fleet in Seattle.

With the adoption of ship classification symbols on July 17, 1920, Oregon was redesignated BB-3. In 1921, a movement began to preserve the battleship as a historic and sentimental relic in Oregon.

In accordance with the Washington Naval Treaty, Oregon was rendered incapable of further warlike service on January 4, 1924, and retained as a naval relic, classified as "unclassified." In June 1925, she was loaned to the State of Oregon, restored, and moored in Portland as a floating monument and museum.

Redesignated IX-22 on February 17, 1941, Oregon's scrap value was deemed vital to the World War II effort. She was struck from the Navy List on November 2, 1942, and sold on December 7. Towed to Kalama, Washington, for dismantling in March, the Navy requested that scrapping halt at the main deck. She was used as a storage hulk or breakwater in the reconquest of Guam, loaded with dynamite and ammunition, and towed to the island by July 1944.

The hulk remained at Guam for several years. During a typhoon on November 14-15, 1948, she broke free and drifted to sea. Located 500 miles southeast of Guam on December 8, she was towed back and sold on March 15, 1956, to the Massey Supply Corp., resold to the Iwai Sanggo Co., towed to Kawasaki, Japan, and scrapped.

BB-4 • USS IOWA

(BB-4: dp. 11,346; l. 360'; b. 72' 2"; dr. 24'; s. 17 k.; cpl. 727; a. 4 12", 8 8", 6 4", 20 6-pdrs., 4 1-pdrs., 24 14" tt)

The second Iowa (BB-4) was laid down by William Cramp & Sons, Philadelphia, on 5 August 1893, launched on 28 March 1896; sponsored by Miss M. L. Drake, daughter of the governor of Iowa; and commissioned on 16 June 1897, Captain W. T. Sampson in command.

After a shakedown off the Atlantic Coast, Iowa was assigned to the Atlantic Fleet and was ordered to blockade duty on 28 May 1898, off Santiago de Cuba. On 3 July 1898, she was the first to sight the Spanish ships approaching and fired the first shot in the Battle of Santiago de Cuba. In a 20-minute battle with Spanish cruisers Maria Teresa (flagship) and Oquendo, her effective fire set both ships aflame and drove them onto the beach. Iowa, continuing the battle in company with the converted yacht Gloucester, sank the Spanish

destroyer Pluton and so damaged destroyer Furor that she ran aground. Iowa then turned her attention to the Spanish cruiser Vizcaya which she pursued until Vizcaya ran aground. Upon the conclusion of the battle, Iowa received on board Spanish Admiral Cervera and the officers and crews of the Vizcaya, Furor, and Pluton.

After the Battle of Santiago, Iowa left Cuban waters for New York, arriving on 20 August 1898. On 12 October 1898, she departed for duty in the Pacific, sailed around Cape Horn, and arrived in San Francisco on 7 February 1899. The battleship then steamed to Bremerton, Wash., where she entered dry dock on 11 June 1899. After repairs, Iowa served in the Pacific Squadron for 2½ years, conducting training cruises, drills, and target practice. Iowa left the Pacific early in February 1902 to become the flagship of the South Atlantic Squadron. She sailed for New York on 12 February 1903 and decommissioned on 30 June 1903.

Iowa recommissioned on 23 December 1903 and joined the North Atlantic Squadron. She participated in the John Paul Jones Commemoration ceremonies on 30 June 1905. Iowa remained in the North Atlantic until she was placed in reserve on 6 July 1907. She decommissioned at Philadelphia on 23 July 1908.

Iowa recommissioned on 2 May 1910 and served as a training ship and as a component of the Atlantic Reserve Fleet. During the next four years, she made a number of training cruises to Northern Europe and participated in the Naval Review at Philadelphia from 10 to 15 October 1912. She decommissioned at Philadelphia Navy Yard on 27 May 1914. At the outbreak of the First World War, Iowa was placed in limited commission on 23 April 1917. After serving as a Receiving Ship at Philadelphia for six months, she was sent to Hampton Roads, Va., and remained there for the duration of the war, training men for other ships of the Fleet, and doing guard duty at the entrance to Chesapeake Bay. She decommissioned for the final time on 31 March 1919.

On 30 April 1919, Iowa was renamed Coast Battleship No. 4, and was the first radio-controlled target ship to be used in a fleet exercise. She was sunk on 23 March 1923 in Panama Bay by a salvo of 14-inch shells.

BB-5 • USS KEARSARGE

(BB-5: displacement 11,540; length 375 feet 4 inches; beam 72 feet 3 inches; draft 23 feet 6 inches; speed 16 knots; complement 553; armament 4 13-inch guns, 4 8-inch guns, 14 5-inch guns, 20 6-pounders, 8 1-pounders, 4 .30 caliber machine guns)

The second ship named Kearsarge, named by an act of Congress to commemorate the famed steam sloop-of-war, was launched on 24 March 1898 by the Newport News Shipbuilding Co., Newport News, Va.; sponsored by Mrs. Herbert Winslow, daughter-in-law of Kearsarge commander, Captain John A. Winslow, during her famous battle with Alabama, and commissioned on 20 February 1900, with Captain William M. Folger in command.

Kearsarge became the flagship of the North Atlantic Station, cruising down the Atlantic seaboard and in the Caribbean. From 3 June 1903 to 26 July 1903, she briefly served as flagship of the European Squadron while on a cruise that took her first to Kiel, Germany. She was visited by the German Emperor on 25 June 1903 and by the Prince of Wales on 13 July. She returned to Bar Harbor, Maine, on 26 July 1903 and resumed duties as flagship of the North Atlantic Fleet. She sailed from New York on 1 December 1903 for Guantanamo Bay, Cuba, where, on 10 December, the United States took formal possession of the Guantanamo Naval Reservation.

Following maneuvers in the Caribbean, she led the North Atlantic Battleship Squadron to Lisbon where she entertained the King of Portugal on 11 June 1904. She next steamed to Phaleron Bay, Greece, where she celebrated the Fourth of July with the King and Prince Andrew, Princess Alice of Greece. The squadron paid goodwill calls at Corfu, Trieste, and Fiume before returning to Newport, R.I., on 29 August 1904.

Kearsarge remained flagship of the North Atlantic Fleet until relieved on 31 March by the battleship Maine but continued operations with the fleet. During target practice off Cape Cruz, Cuba, on 13 April 1906, an accidental ignition of a powder charge in a 13-inch gun killed two officers and eight men. Four men were seriously injured. Attached to the 2nd Squadron, 4th Division, she sailed on 16 December 1907 with the "Great White Fleet" of battleships, sent around the world by President Theodore Roosevelt. She sailed from Hampton Roads around the coasts of South America to the western seaboard, then to Hawaii, Australia, New Zealand, the Philippines, and Japan. From there, Kearsarge proceeded to Ceylon, transited the Suez Canal, and visited ports of the Mediterranean before returning to the eastern seaboard of the United States. President Theodore Roosevelt reviewed the Fleet as it passed into Hampton Roads on 22 February 1909, having completed a world cruise of overwhelming success, showing the flag and spreading goodwill. This dramatic gesture impressed the world with the power of the U.S. Navy.

Kearsarge decommissioned in the Philadelphia Navy Yard on 4 September 1909 for modernization. She recommissioned on 23 June 1915 for operations along the Atlantic coast until 17 September when she departed Philadelphia to land a detachment of marines at Vera Cruz, Mexico. She remained off Vera Cruz from 28 September 1915 to 5 January 1916, then carried the marines to New Orleans before joining the Atlantic Reserve Fleet on 4 February 1916 at Philadelphia. She trained Massachusetts and Maine State Naval Militia until America entered World War I, then trained thousands of armed guard crews as well as naval engineers in waters along the East Coast ranging from Boston to Pensacola. On the evening of 18 August 1918, Kearsarge rescued 26 survivors of the Norwegian Bark Nordhav, which had been sunk by German Submarine U 117. The survivors were landed in Boston.

Kearsarge continued as an engineering training ship until 29 May 1919 when she embarked Naval Academy Midshipmen for training in the West Indies. The midshipmen were debarked at Annapolis on 29 August and Kearsarge proceeded to the Philadelphia Navy Yard, where she decommissioned on 10 May 1920 for conversion to a crane ship and a new career. She was designated AB-1 on 5 August 1920.

In place of military trappings, Kearsarge received an immense revolving crane with a rated lifting capacity of 250 tons, as well as hull "blisters," which gave her more stability. The 10,000-ton crane ship rendered invaluable service for the next 20 years. One of many accomplishments was the raising of the sunken submarine Squalus off the New Hampshire coast. On 6 November 1941, she was designated Crane Ship No. 1, giving up her illustrious name, which was assigned to a mighty aircraft carrier. But she continued her yeoman service and made many contributions to the American victories of World War II. She handled guns, turrets, armor, and other heavy lifts for new battleships such as Indiana and Alabama, cruisers Savannah and Chicago, and guns on the veteran battleship Pennsylvania.

In 1945, the crane ship was towed to the San Francisco Naval Shipyard where she assisted in the construction of carriers Hornet, Boxer, and Saratoga. She departed the West Coast in 1948 to finish her career in the Boston Naval Shipyard. Joe McDonald, master rigger, described her as "a big gray hulk of a thing" which was "pulled around by two or three tugs" on the job; "But the old girl has brought millions of dollars' worth of business to Boston. Without her, we would never have been able to do many of the big jobs that cost millions of dollars." As one example, he recalled that the former battleship lifted a gantry crane intact at the South Boston Naval Drydocks and transported it to Charleston where she placed it on crane tracks to be driven away. As Crane Ship No. 1, her name was struck from the Navy List on 22 June 1955. She was sold for scrapping on 9 August 1955.

BB-6 • USS KENTUCKY

BB-6: displacement 11,520; length 375'4"; beam 72'2½"; speed 16.9 knots; complement 554; armament 4 13", 4 8", 14 5", 20 6-pounders, 8 1-pounders, 4 .30 caliber machine guns, 4 18" torpedo tubes; class Kearsarge)

Kentucky (BB-6) was launched on 24 March 1898 by Newport News Shipbuilding & Dry Dock Co., Newport News, Va.; sponsored by Miss Christine Bradley, daughter of Governor William Bradley of Kentucky; and commissioned on 15 May 1900, with Captain Colby M. Chester in command.

After fitting out at the New York Navy Yard during the summer, Kentucky sailed on 25 October 1900 for the Far East via Gibraltar and the Suez Canal. She joined the other American ships on the Asiatic Station at Manila on 3 February 1901 and six days later sailed for Hong Kong, where she became the flagship of the Southern Squadron under Rear Admiral Louis Kempff on 23

March. Throughout the following year, the battleship led her squadron as it watched over American interests in the Far East, visiting principal ports of China and Japan including Chefoo, Taku, Nanking, Woosung, Hong Kong, Amoy, Nagasaki, Kobe, and Yokohama.

Rear Admiral Frank F. Wildes also selected Kentucky as his flagship upon relieving Admiral Kempff on 1 March 1902, but he transferred his flag to Rainbow on 7 April. Rear Admiral Robley D. Evans, Commander in Chief, Asiatic Fleet, chose Kentucky as his flagship at Yokohama on 4 November; and he continued to direct American naval operations in the Far East from her until she sailed from Manila for home on 13 March 1904. After retracing her steps through the Suez Canal and the Strait of Gibraltar, she arrived in New York on 23 May.

Upon completing overhaul in the New York Navy Yard on 26 October, Kentucky devoted the following year to tactics and maneuvers off the Atlantic coast with the North Atlantic Fleet. The battleship joined the welcome of the British Squadron at Annapolis and New York in the fall of 1905 and then cruised along the eastern seaboard until 23 September 1906. On that day off Provincetown, she embarked marines from Maine, Missouri, and Kearsarge and landed them at Havana on 1 October to protect American lives and property during the Cuban Insurrection. She stood by to support forces ashore until 9 October before resuming battle practice and tactics in the North Atlantic.

Kentucky visited Norfolk on 15 April 1907 to attend the Jamestown Exposition; and, after more exercises off the New England coast, she returned to Hampton Roads to join the "Great White Fleet" of 16 battleships for a world cruise that brought great prestige and honor to the Navy and the Nation. Rear Admiral Evans, Kentucky's former Flag Officer, commanded the fleet as it circumnavigated the globe, receiving warm and enthusiastic welcomes at each port of call. As the famous voyage got underway from Hampton Roads on 16 December, Kentucky passed in review before President Roosevelt as a unit in the 2nd Squadron. After calling at Trinidad and Rio de Janeiro, the warships passed in open order through the Straits of Magellan to visit Punta Arenas and Valparaiso, Chile. A stop at Callao Bay, Peru, was followed by a month of target practice out of Magdalena Bay, Mexico. The fleet reached San Diego on 14 April 1908 and moved on to San Francisco on 7 May. Exactly two months later, the spotless warships sortied through the Golden Gate and sailed for Honolulu. From Hawaii, they set course for Auckland, New Zealand, arriving on 9 August. The fleet made Sydney on 20 August and, after a week of warm and cordial hospitality, sailed for Melbourne.

Kentucky departed Albany, Australia, on 10 September for ports in the Philippine Islands, Japan, China, and Ceylon before transiting the Suez Canal. She departed Port Said on 8 January 1909 to visit Tripoli and Algiers with the 4th Division before reforming with the fleet at Gibraltar. Underway for home on 6 February, she again passed in review before President Roosevelt upon entering Hampton Roads on 22 February, ending a widely acclaimed voyage of goodwill in which she and her sister ships subtly but effectively demonstrated American strength to the world.

After local operations and repairs at the Philadelphia Navy Yard, Kentucky decommissioned at Norfolk on 28 August 1909. She recommissioned in the 2nd Reserve on 4 June 1912 but, save for a run to

New York, did not operate at sea before being placed in ordinary in the Philadelphia Navy Yard on 31 May 1913.

The veteran battleship recommissioned at Philadelphia on 23 June 1915 and sailed on 3 July to train New York militia in a cruise from Long Island to ports in New England and Chesapeake Bay. She debarked the militia at New York and sailed to Portland to embark Maine militia for a training cruise. Returning to Philadelphia on 31 August, she sailed on 11 September for the coast of Mexico to watch over American interests during the unrest caused by the Mexican Revolution. She reached Vera Cruz on 28 September 1915; and, but for a visit to New Orleans for Mardi Gras in March 1916, she remained on patrol off the Mexican coast until 2 June 1916.

The battleship called at Guantanamo Bay and Santo Domingo en route home to Philadelphia, where she arrived on 18 June. Following maneuvers and tactics ranging north to Newport during the summer, Kentucky arrived in New York on 2 October and remained in the North River until the end of the year. She entered the New York Naval Shipyard for repairs on 1 January 1917 and was still there when the United States entered World War I. She arrived in Yorktown, Va., on 2 May for duty as a training ship and trained recruits on cruises in Chesapeake Bay and along the Atlantic coast as far north as Long Island Sound. When the Armistice was signed on 11 November 1918, she was training her 15th group of recruits, having already trained several thousand men for service in ships of the war-expanded Navy.

Kentucky entered the Boston Navy Yard on 20 December for overhaul. She sailed on 18 March 1919 for refresher training out of Guantanamo Bay and then participated in fleet maneuvers and exercises ranging north from Norfolk to the New England coast. She arrived in Annapolis on 29 May to embark midshipmen and got underway on 9 June for a summer practice cruise that took her to Cuba, the Virgin Islands, Panama, New York, Boston, and Provincetown. She returned to Annapolis on 27 August to debark her midshipmen and entered the Philadelphia Navy Yard on 30 August. She remained there until decommissioning on 29 May 1920. Kentucky was sold to Dravo Construction Co., Pittsburgh, Pa., for scrapping on 23 January 1924 in compliance with U.S. commitments under the Washington Treaty which limited naval armaments.

BB-7 • USS ILLINOIS

(BB-7: dp. 11,565 (n.), l. 374'10"; b. 72'5", dr. 25'0" (f.) (aft) s. 16 k., colt 536, a. 4 13" 14 6", 16 6-pars., 4 1-pars.4 .30-cal. mg., 4 18" tt.; cl. Illinois)

Illinois (BB-7) was laid down on 10 February 1897 by the Newport News Shipbuilding & Dry Dock Co., Newport News, Va., launched on 4 October 1898; sponsored by Miss Nancy Leiter; and commissioned on 16 September 1901 with Captain G.A. Converse in command.

After shakedown and training in Chesapeake Bay, the new battleship sailed on 20 November 1901 for Algiers, La., where she was used to test a new floating dry dock. She returned to Newport News in January 1902, and from 15 to 28 February, Illinois served as the flagship for Rear Admiral R.D. Evans during the reception for Prince Henry of Prussia. Bearing the flag of Rear Admiral A.S. Crowninshield, the battleship departed New York on 30 April 1902 and arrived in Naples on 18 May, where the Admiral took command of the European Squadron.

Illinois carried out training and ceremonial duties until 14 July 1902, when she grounded in the harbor of Christiana, Norway, and had to return to England for repairs. She remained at Chatham until 1 September 1902, then proceeded to the Mediterranean and South Atlantic for fleet maneuvers.

Illinois was detached from the European Squadron on 10 January 1903 and assigned to the North Atlantic. She engaged in fleet maneuvers, gunnery, and seamanship training, and ceremonial operations until December 1907, when she joined the Great White Fleet for the cruise around the world. Following a Presidential review, the mighty battleships sailed from Hampton Roads on their important voyage. The Atlantic Fleet joined the Pacific Fleet on 8 May 1908 in San Francisco Bay, and after a review by the Secretary of the Navy, the combined fleets continued their cruise. The ships visited Australia, Japan, Ceylon, and other countries, arriving in Suez on 3 January 1909. At Suez, word of the Sicilian earthquake sent Illinois, Connecticut, and Culgoa to Messina. After rendering valuable aid to victims of the disaster, the ships rejoined the fleet, returning to Hampton Roads on 22 February 1909. President Roosevelt reviewed the fleet as it arrived, having given the world a graphic demonstration of America's naval might. Illinois was decommissioned at Boston on 4 August 1909.

The battleship was placed in reserve commission on 15 April 1912 and recommissioned on 2 November 1912, in time to take part in winter fleet exercises and battle maneuvers with the Atlantic Fleet. During the summers of 1913 and 1914, Illinois made training cruises to Europe with midshipmen. In 1919, the ship was again laid up at the Philadelphia Navy Yard and was later loaned to the State of New York on 23 October 1921 for use by the Naval Militia.

Excluded from further use as a warship by the terms of the Washington Treaty, Illinois was fitted out as a floating armory at the New York Navy Yard in 1924 and was assigned to the New York Naval Reserve. She remained there for more than 30 years, though reclassified as IX-15 on 8 January 1941 and renamed Prairie State to allow her name to be assigned to a projected new battleship. During World War II, she served as a U.S. Naval Reserve Midshipmen Training School in New York. Following the war, she was retained on loan to the State as quarters for a Naval Reserve unit until 31 December 1955.

Prairie State, after over 50 years of useful service to the Navy and the Nation, was towed to Baltimore and sold for scrap to Bethlehem Steel Co. on 18 May 1956.

BB-8 • USS ALABAMA

(BB- 8: dp. 11,565 (n.), 1. 374'10"; b. 72'5", dr. 25'0" (f.) (aft) s. 16 k., colt 536, a. 4 13" 14 6", 16 6-pars., 4 1-pars.4 .30-cal. mg., 4 18" tt.; cl. Illinois).

The second name ship Alabama (Battleship No. 8) was laid down on December 1, 1896, at Philadelphia, Pa., by the William Cramp and Sons Ship and Engine Building Co., launched on May 18, 1898, sponsored by Miss Mary Morgan, daughter of the Honorable John T. Morgan, United States Senator from Georgia; and commissioned on October 16, 1900, Capt. Willard H. Brownson in command.

Though assigned to the North Atlantic Station, Alabama did not begin operations with that unit until early the following year. The warship remained in Philadelphia until December 13, when she went underway for the brief trip to New York. She stayed in New York through the New Year and until the latter part of January 1901. Finally, on January 27, the battleship headed south for winter exercises with the Fleet at the drill grounds in the Gulf of Mexico near Pensacola, Fla. Alabama's Navy career began in earnest with her arrival in the gulf early in February. With a single exception in 1904, each year from 1901 to 1907, she conducted Fleet exercises and gunnery drills in the Gulf of Mexico and the West Indies in the wintertime before returning north for repairs and operations off the northeastern coast during the summer and autumn. The exception came in the spring of 1904 after the conclusion of winter maneuvers when she departed Pensacola in company with Kearsarge (Battleship No. 5), Maine (Battleship No. 10), Iowa (Battleship No. 4), Olympia (Cruiser No. 6), Baltimore (Cruiser No. 3), and Cleveland (Cruiser No. 19) on a voyage to Portugal and the Mediterranean. After a ceremonial visit to Lisbon honoring the entrance of the Infante into the Portuguese naval school, Alabama and the other three battleships cruised the Mediterranean until mid-August. Returning by way of the Azores, she and her traveling company- arrived in Newport, R. I., on August 29. Late in September, the warship entered the League Island Navy Yard for repairs. Early in December, Alabama left the yard and resumed cruising with the North Atlantic Fleet.

Near the end of 1907, the battleship set out on a special mission. On December 16, 1907, she stood out of Hampton Roads in company with what became known as the Great White Fleet. Alabama accompanied the Fleet on its voyage around the South American continent as far as San Francisco. On May 18, 1908, when the bulk of the Fleet headed north to visit the Pacific Northwest, she remained in San Francisco for repairs at the Mare Island Navy Yard. As a consequence, the warship did not participate in the celebrated visit to Japan. Instead, Alabama and Maine departed San Francisco on June 8 to complete their own, more direct circumnavigation of the globe. Steaming by way of Honolulu and Guam, the two battleships arrived in Manila, Philippines, on July 20. In August, they visited Singapore and Colombo on the island of Ceylon. From Colombo, the two battleships made their way, via Aden on the Arabian Peninsula, to the Suez Canal. Through the canal early in September, Alabama and Maine made an expeditious transit of the Mediterranean Sea, pausing only at Naples at mid-month. Following a port call at Gibraltar, they embarked upon the Atlantic passage on October 4. They made one stop in the Azores on their way across the Atlantic. On October 19, as they neared the end of their long voyage, the two battleships Parted company. Maine headed for Portsmouth, N.H., and Alabama steered for New York. Both reached their destinations on the 20th.

Alabama was placed in reserve at New York on November 3, 1908. Though she remained inactive in New York, the battleship was not decommissioned until August 17, 1909. The warship underwent an extensive overhaul that lasted until the early part of 1912. On April 17, 1912, she was placed in commission, second reserve, at New York, Comdr. Charles F. Preston in command. At that point, she became an element of the newly established Atlantic Reserve Fleet. According to that concept, the Navy organized a unit that comprised nine of the older battleships as well as Brooklyn (Armored Cruiser No. 3), Columbia (Cruiser No. 12), and Minneapolis (Cruiser No. 13) for the purpose of keeping those ships constantly ready for active service using the fiscal expedient of severely reduced complements that could be filled out rapidly by naval militiamen and volunteers in an emergency. The unit as a whole possessed enough officers and men to take two or three of the ships to sea on a rotating basis to test their material readiness and to exercise the sailors at drill.

Alabama was placed in full commission on July 25, 1912, and operated with the Atlantic Fleet off the New England coast through the summer. She was returned

to reserve status—in commission, first reserve—in New York on September 10, 1912. Late in the spring of 1913, the Navy added a new dimension to the concept of the Atlantic Reserve Fleet by having the warships of that unit embark detachments of the various state naval militias for training afloat in a manner similar in many respects to the contemporary Navy's selected reserve program. During the summer of 1913, Alabama cruised along the East Coast and made two round-trip voyages to Bermuda to train naval militiamen from Maryland, the District of Columbia, New York, Rhode Island, Maine, North Carolina, and Indiana. She ended her last training cruise of the year in Philadelphia on September 2. The battleship was placed in ordinary on October 31, 1913, and in reserve on July 1, 1914.

Though still in commission, she passed the next 30 months in relative inactivity with the Reserve Force, Atlantic Fleet, at Philadelphia. America's shift toward belligerency in World War I, however, brought Alabama out of the doldrums of the peacetime reserve at the beginning of 1917. On January 22, she became a receiving ship in Philadelphia, embarking on drafts of recruits for training. In mid-March, the battleship moved south to the lower reaches of the Chesapeake Bay and began transforming landsmen into sailors. She took a brief respite from her rigorous training schedule on April 6, 1917, for the announcement of the United States' declaration of war on the Central Powers. Two days later, Alabama became the flagship of Division 1 Atlantic Fleet. For the remainder of World War I, the warship conducted recruit training missions in the lower Chesapeake Bay and in the coastal waters of the Atlantic seaboard, though she made one visit to the Gulf of Mexico in late June and early July of 1918.

After the armistice on November 11, 1918, her recruit training duties continued but began to diminish somewhat in intensity. During February and March of 1919, the battleship steamed south to the West Indies for winter maneuvers. She returned to Philadelphia in mid-April for routine repairs before heading for Annapolis to embark Naval Academy midshipmen for their summer training cruise. On May 28 and 29, Alabama made the short trip from Philadelphia to Annapolis. She left Annapolis on June 9 with 184 midshipmen embarked. During the first part of the cruise, Alabama visited the West Indies and made a trip through the Panama Canal and back. In mid-July, she voyaged to New York and the New England coast. August saw her return south for maneuvers at the drill grounds. Alabama disembarked the midshipmen at Annapolis at the end of August and returned to Philadelphia.

After more than nine months at Philadelphia lingering in a sort of naval purgatory, the battleship was finally decommissioned on May 7, 1920. On September 15, 1921, Alabama was transferred to the War Department to be used as a target, and her name was struck from the Navy list. Subjected to aerial bombing tests in Chesapeake Bay by planes of the Army Air Service, the warship sank in shallow water on September 27, 1921. On March 19, 1924, her sunken hulk was sold for scrap.

Alabama is a 69-foot motorboat built in 1906 at South Boston, Mass., by George Lawley and Sons. It was inspected by the Navy in the summer of 1917. Records indicate that on July 25, 1917, the Navy concluded an agreement with her owners, the American and British Manufacturing Co. Bridgeport, Conn., for possible future acquisition of the boat. By the terms of that agreement, Alabama was assigned the designation SP-1052 and was "enrolled in the Naval Coast Defense Reserve." All indications are, however, that Alabama never saw actual naval service, possibly remaining "enrolled" in a reserve capacity, since she does not appear on contemporary lists of commandeered, chartered, or leased small craft actually used by the Navy during World War I.'

BB-9 • USS WISCONSIN

(BB-9: dp. 11,564 (n.); l. 373'10", b. 72'2.5", dr. 23'8.1" (mean); s. 16 k., cpl. 531; a. 4 13", 14 6", 16 6-pers., 6 1-pdrs., 4 .30-cal. mg.;cl. Illinois)

First Wisconsin (Battleship No. 9) was laid down on February 9, 1897, at San Francisco, Calif., by the Union Iron Works, launched on November 26, 1898, sponsored by Miss Elizabeth Stephenson, the daughter of Senator Isase Stephenson of Marinette, Wis., and commissioned on February 4, 1901, Capt. George C. Reiter in command.

Departing San Francisco on March 12, 1901, Wisconsin conducted general drills and exercises at Magdalena Bay, Mexico, from March 17 to April 11 before she returned to San Francisco on April 15 to be dry docked for repairs. Upon completion of that work, Wisconsin headed north along the western seaboard, departing San Francisco on May 28 and reaching Port Orchard Wash., on June 1. She remained there for nine days before heading back toward San Francisco.

She next made a voyage—in company with the battleships Oregon and Iowa, the cruiser Philadelphia, and the torpedo-boat destroyer Farragut to the Pacific Northwest, reaching Port Angeles, Wash., on June 29. She then shifted to Port Whatcom, Wash., on July 2 and participated in the 4th of July observances there before she returned to Port Angeles the following day to resume her scheduled drills and exercises. Those evolutions kept the ship occupied through mid-July.

Following repairs and alterations at the Puget Sound Navy Yard, Bremerton, Wash., from July 23 to October 14, Wisconsin sailed for the middle and southern reaches of the Pacific, reaching Honolulu, Hawaii, on October 23. After coaling there, the battleship then got underway for Samoa on the 26th and exercised her main and secondary batteries en route to her destination.

Reaching the naval station at Tutuila on November 5, Wisconsin remained in that vicinity, along with the collier Abarenda and the hospital ship Solace, for a little over two weeks. Shifting to Apia—the scene of the disastrous hurricane of 1888—Wisconsin hosted the Governor of German Samoa before the man-of-war departed that port on the 21st, bound—via Hawaii—for the coastal waters of Central and South America.

Wisconsin reached Acapulco on Christmas Day, 1901, and remained in port for three days. After coaling, the man-of-war twice visited Callao, Peru, and also called at Valparaiso, Chile, before she returned to Acapulco on February 26, 1902.

Wisconsin exercised in Mexican waters—at Pichilinque Bay and Magdalena Bay—from 5 to March 22, carrying out an intensive and varied slate of exercises that included small arms drills, day and night main battery target practices, and landing force maneuvers. She conducted further drills of various kinds as she proceeded up the west coast, touching Coronado, San Francisco, and Port Angeles before she reached the Pu~et Sound Navy Yard on June 4.

The battleship underwent repairs and alterations until August 11. She then conducted gunnery exercises off Tacoma and Seattle, Wash., before she returned to the Puget Sound Navy Yard on August 29 for further work. : She remained there until September 12, when she sailed for San Francisco en route to Panama.

Wisconsin—as flagship, Pacific Squadron—with Rear Admiral Silas Casey embarked and arrived at Panama, Colombia, on September 30, 1902, to protect American interests and to preserve the integrity of transit across the isthmus. Casey offered his services as a mediator in the crisis that had lasted for three years and invited leaders of both factions—conservatives and liberals— to meet onboard Wisconsin. Over succeeding weeks through October and into November, prolonged negotiations ensued. Ultimately, however, the warring sides came to an agreement and signed a treaty on November 21, 1902. The accord came to be honored in Colombian circles as "The Peace of Wisconsin." When Rear Admiral Henry Glass, Admiral Casey's successor as Commander in Chief of the Pacific Squadron, wrote his report to the Secretary of the Navy for the fiscal year 1903, he lauded his predecessor's diplomatic services during the Panama crisis. "The final

settlement of the revolutionary disturbance," Glass wrote approvingly, "was largely due to his efforts."

Her task completed, the battleship departed Panama's waters on November 22 and arrived at San Francisco on December 5 to prepare for gunnery exercises. Four days later, Rear Admiral Casey shifted his flag to the armored cruiser New York, thus releasing Wisconsin from flagship duties for the Pacific Squadron. The battleship consequently carried out her firings until December 17, when she sailed for Bremerton. Reaching the Puget Sound Navy Yard five days before Christmas of 1902, Wisconsin then underwent repairs and alterations until May 13, 1903, when she sailed for the Asiatic Station.

Proceeding via Honolulu, Wisconsin, arrived at Yokohama, Japan, on June 12, with Rear Admiral Yates Stirling embarked; three days later, Rear Admiral Stirling exchanged flagships with Rear Admiral P. H. Cooper, who broke his two-starred flag at Wisconsin's main as Commander of the Asiatic Fleet's Northern Squadron while Admiral Stirling hoisted his in the tender Rainbow.

Wisconsin operated in the Far East with the Asiatic Fleet over the next three years before she returned to the United States in the autumn of 1906. She followed a normal routine of operations in the northern latitudes of the station—China and Japan—in the summer months because of the oppressive heat of the Philippine Islands that time of year but in the Philippine Archipelago in the winter. She touched at ports in Japan and China, including Kobe, Yokohama, Nagasaki, and Yokosuka; Amoy, Shanghai, Chefoo, Nanking, and Taku. In addition, she cruised the Yangtze River (as far as Nanking), the Inland Sea, and Nimrod Sound. The battleship conducted assigned fleet maneuvers and exercises off the Chinese and Philippine coasts, intervening in those evolutions with regular periods of in-port upkeep and repairs. During that time, she served as the Asiatic Fleet flagship, wearing the flag of Rear Admiral Cooper.

The battleship departed Yokohama on September 20 and, after calling at Honolulu en route between 3 and 8 October, arrived at San Francisco on the 18th. After seven days' stay at that port, she headed up the west coast and reached the Puget Sound Navy Yard on October 28. She was decommissioned there on November 15, 1906.

Recommissioned on April 1, 1908, Capt. Henry Morrell, in command, Wisconsin, was fitted out at the Puget Sound Navy Yard until the end of April. After shifting to Port Angeles from April 30 to May 2, the battleship proceeded down the western seaboard and reached San Francisco on May 6 to participate in a fleet review at that port. She subsequently returned to Puget Sound to complete the installation of her fire control equipment between May 21 and June 22.

Soon thereafter, Wisconsin retraced her southward course, returning to San Francisco in early July. There, she joined the battleships of the Atlantic Fleet in setting out on the transpacific leg of the momentous circumnavigation of the globe. The cruise of the "Great White Fleet" served as a pointed reminder to Japan of the power of the United States—a dramatic gesture made by President Theodore Roosevelt as signal evidence of his "big stick" policy. Wisconsin, during the course of her part of the voyage, called at ports in New Zealand, Australia, the Philippines, Japan, China, Ceylon, and Egypt, transited the Suez Canal, visited Malta, Algiers, and Gibraltar before arriving in Hampton Roads on Washington's Birthday, 1909, and passing in review there before President Roosevelt. The epic voyage had confounded the doomsayers and critics, having been accomplished without any serious incidents or mishaps.

Wisconsin departed from the Tidewater area on March 6 and arrived at the Portsmouth (N.H.) Navy Yard three days later. The pre-dreadnought battleship there underwent repairs and alterations until June 23, giving it her bright "white and spar color" and donning a more businesslike gray. The man-of-war joined the Atlantic Fleet in Hampton Roads at the end of June, but she remained in those waters only a short time, for she sailed north to Portland, Maine, arriving there on July 2 in time to take part in the 4th of July festivities in that port.

The battleship next headed down the eastern seaboard, cruising off Rockport and Provincetown, Mass., before she returned, with the fleet, to Hampton Roads on August 6. Over the ensuing weeks, Wisconsin fired target practices in the southern drill grounds, off the Virginia capes, breaking those underway periods with upkeep in Hampton Roads.

Wisconsin steamed with the fleet to New York City—where she anchored in the North River to take part in the Hudson-Fulton celebrations between September 22 and October 5—before she underwent repairs at the Portsmouth (N.H.) Navy Yard from October 7 to November 28. She then dropped down to Newport, B.I., upon the conclusion of that yard period, picking updrafts of men for transportation to the Atlantic Fleet at Hampton Roads.

Wisconsin operated with the fleet off the Virginia capes through mid-December before she headed for New York for the Christmas holidays in port. Subsequently cruising to Cuban waters in early January 1910, the battleship operated out of Guantanamo Bay for a little over two months, from January 12 to March 19.

The pre-dreadnought battleship then visited Tompkinsville, N.Y., and New Orleans, La., before she discharged ammunition at New York City on April 22. Later that spring, 1910, she moved to the Portsmouth (N.H.) Navy Yard, where she was placed in reserve. She was moved to Philadelphia in April 1912 and, that autumn, took part in a naval review off Yonkers, New York, before resuming her reserve status with the Atlantic Reserve Fleet. Placed "in ordinary" on October 31, 1913, Wisconsin remained in that status until she joined the Naval Academy Practice Squadron in the spring of 1915, assuming training duties along with the battleships Missouri and Ohio. With that group, she became the third battleship to transit the Panama Canal, making that trip in mid-July 1915 en route to the west coast of the United States with her embarked officers-to-be.

Wisconsin discharged her duties as a midshipman's training ship into 1917 and was moored at the Philadelphia Navy Yard on April 6 of that year, when she received word that the United States had declared war on Germany. Two days later, members of the Naval Militia began reporting on board the battleship for quarters and subsistence.

On April 23, Wisconsin, Missouri, and Ohio were placed in full commission and assigned to the Coast

Battleship Patrol Squadron. Within two weeks, on May 2, Comdr. (later Admiral) David F. Sellers reported on board and took command. Four days later, the battleship got underway for the Virginia Capes, and she arrived at Yorktown, VA., on the 7th.

From early May through early August, Wisconsin operated as an engineering school ship on training cruises in the Chesapeake Bay-York River area. She trained recruits as oilers, watertenders, and firemen—who, when qualified, were assigned to the formerly interned merchantmen of the enemy taken over by the United States upon the declaration of war, as well as to submarine chasers and the merchant vessels then building in American yards.

Wisconsin then maneuvered and exercised in company with the battleships Kearsarge, Alabama, Illinois, Kentucky, Ohio, Missouri, and Maine between 13 and 19 August, en route to Port Jefferson, L.I. Over the ensuing weeks, Wisconsin continued training and tactical maneuvers based on Port Jefferson, making various training cruises into Long Island Sound.

She subsequently returned to the York River region early in October and resumed her training activities in that locale, operating primarily in the Chesapeake Bay area. Wisconsin continued that duty into the spring of 1918, interrupting her training evolutions between October 30 and December 18, 1917, for repairs at the Philadelphia Navy Yard.

After another stint of repairs at Philadelphia from May 13 to June 3, 1918, Wisconsin got underway for a cruise to Annapolis but, after passing the Brandywine Shoal Light, received orders to stick close to shore. Those orders were later modified to send Wisconsin Up the Delaware River as far as Bombay Hook since an enemy submarine was active off Cape Henlopen. Postwar examination of German records would show that U-l 51—reportedly the first of six enemy submarines to come to the eastern seaboard in 1918—sank three schooners on May 23 and other ships over ensuing days.

Getting underway again on June 6, Wisconsin arrived at Annapolis on the following day. On the next day, the battleship embarked 175 3d class midshipmen and got underway for the York River. The ship conducted training evolutions in the Chesapeake Bay region until August 29, when she returned to Annapolis and disembarked midshipmen. Underway for Yorktown on the 30th, Wisconsin there. She embarked on the task of training 217 men as firemen, water tenders, engineers, steersmen, and signalmen, resumed her training duties, and continued the task through the signing of the armistice on November 11.

She completed her training activities on December 20, sailed north, and reached New York City three days before Christmas. Wisconsin was among the ships reviewed by Secretary of the Navy Josephus Daniels from the deck of the yacht Mayflower and by Assistant Secretary of the Navy Franklin D. Roosevelt from Azlec (SP-690) on the day after Christmas, December 26.

Wisconsin cruised with the fleet in Cuban waters that winter and, in the summer of 1919, made a midshipman training cruise to the Caribbean.

Placed out of commission on May 15, 1920, Wisconsin was reclassified BB-9 on July 17, 1920, while awaiting disposition. She was sold for scrap on January 26, 1922, as a result of the Washington Treaty.

BB-10 • USS MAINE

(Battleship No. 10: Displacement: 12,846 (normal); Length: 393'11"; Beam: 72'3"; Draft: 24'4"; Speed: 18 knots; Complement: 561; Armament: 4 12-inch guns, 16 6-inch guns, 6 3-inch guns, 8 3-pounders, 6 1-pounders, 2 18-inch torpedo tubes; Class: Maine)

The second Maine (Battleship No. 10) was laid down by William Cramp & Sons in Philadelphia, Pennsylvania, on 15 February 1899, exactly one year after the destruction of the first Maine. She was launched on 27 July 1901, sponsored by Miss Mary Preble Anderson, and commissioned at Philadelphia on 29 December 1902, with Captain Eugene H.C. Leutze in command.

From 1903 to 1907, Maine cruised along the Atlantic Coast, south to the West Indies, and completed one cruise to the Mediterranean. On 16 December 1907, she left Hampton Roads with the rest of the Atlantic Fleet en route to the Pacific, where she joined ships of that fleet for a cruise around the world. In company with the Alabama, she traveled to Guam and the Philippines, through the Suez Canal and the Mediterranean, and returned to the Atlantic coast in October 1908, well ahead of the rest of the "Great White Fleet."

Fitted out as the flagship of the 3rd Squadron, Atlantic Fleet, Maine resumed operations along the Atlantic coast and into Caribbean waters over the next several months. She decommissioned at Portsmouth, New Hampshire, on 31 August 1909. Recommissioned on 15 June 1911, Maine operated along the East Coast. During World War I, she trained engineers, armed guard crews, and midshipmen. Following the defeat of the Central Powers, she took part in the review of the fleet in New York on 26 December 1918.

Maine operated with ships of the Atlantic Fleet until 15 May 1920, when she decommissioned at the Philadelphia Navy Yard. Classified as BB-10 on 17 July 1920, she was sold to J.G. Hitner & W.F. Cutler of Philadelphia, Pennsylvania, on 23 January 1922. She was rendered incapable of further warlike service on 17 December 1923 and subsequently broken up and scrapped in accordance with the terms of the Washington Naval Treaty limiting naval armaments.

BB-11 • USS MISSOURI

(BB-11: Displacement: 12,846 (normal); Length: 393'11"; Beam: 72'3"; Draft: 24'4"; Speed: 18 knots; Complement: 561; Armament: 4 12-inch guns, 16 6-inch guns, 6 3-inch guns, 8 3-pounders, 6 1-pounders, 2 18-inch torpedo tubes; Class: Maine)

The third Missouri (BB-11), was laid down by the Newport News Shipbuilding & Drydock Co., Newport News, Virginia, on 7 February 1900. Launched on 28 December 1901, she was sponsored by Mrs. Edson Gallaudet, daughter of Senator Francis Marion Cockrell of Missouri, and commissioned on 1 December 1903, with Captain William S. Cowles in command.

Assigned to the North Atlantic Fleet, Missouri departed Norfolk on 4 February 1904 for trials off the Virginia Capes and fleet operations in the Caribbean. On 13 April, during target practice, a flareback from the port gun in her after turret ignited a powder charge, setting off two others. Although no explosion occurred, the rapid burning of the powder suffocated 36 crew members. Prompt action prevented the loss of the warship, and three crew members were awarded Medals of Honor for extraordinary heroism. After repairs at Newport News, Missouri sailed on 9 June for duty in the Mediterranean, returning to New York on 17 December.

Fleet operations along the East Coast and in the Caribbean during the next few years included providing relief to earthquake victims in Kingston, Jamaica, from 17 to 19 January 1907. In April, she participated in the Jamestown Exposition.

With the "Great White Fleet," Missouri sailed from Hampton Roads on 16 December 1907, passing in review before President Theodore Roosevelt at the start of a world cruise demonstrating American naval might. The fleet visited ports in the Caribbean, along the South American east coast, rounded Cape Horn, and stopped in Peru and Mexico before arriving in San Francisco in

May 1908 for a gala visit. In July, the fleet sailed west to Honolulu, New Zealand, and Australia, reaching Manila on 2 October. After receiving a tumultuous welcome in Yokohama and visiting Amoy, China, the fleet began its homeward journey via Ceylon, the Suez, and ports in the Eastern Mediterranean. Departing Gibraltar on 6 February 1909, the fleet was again reviewed by President Roosevelt upon its triumphant return to Hampton Roads on 22 February, having successfully completed an important diplomatic mission.

Missouri was placed in reserve at Boston on 1 May 1910, recommissioned on 1 June 1911, and resumed East Coast and Caribbean operations with the Atlantic Fleet. In June 1912, she transported marines from New York to Cuba to protect American interests during a rebellion. The following month, she carried midshipmen for training, then decommissioned at Philadelphia on 9 September 1912.

Recommissioned on 16 March 1914, Missouri joined that summer's Naval Academy Practice Squadron's cruise to Italian and English ports, returning to ordinary at Philadelphia on 2 December 1914. She recommissioned on 15 April 1915 for midshipmen

training in the Caribbean and a cruise through the Panama Canal to California ports. She returned to the Reserve Fleet at Philadelphia on 18 October 1915, recommissioned on 2 May 1916, and conducted training along the East Coast and in the Caribbean until placed in ordinary for the winter at Philadelphia.

Upon the United States' entry into World War I, Missouri recommissioned on 23 April 1917, joined the Atlantic Fleet at Yorktown, Virginia, and served as a training ship in the Chesapeake Bay area. On 26 August 1917, Rear Admiral Hugh Rodman hoisted his flag on Missouri as Commander, Division 2, Atlantic Fleet. The warship continued training thousands of recruits in engineering and gunnery for foreign service on warships and as armed guards for merchant vessels.

Following the Armistice, the battleship joined the Cruiser and Transport Force, departing Norfolk on 18 February 1919 on the first of four voyages to Brest to return 3,278 U.S. troops to East Coast ports. Missouri decommissioned at the Philadelphia Navy Yard on 8 September 1919. She was sold to J. G. Hitner and W.F. Cutler of Philadelphia on 26 January 1922 and scrapped in accordance with the treaty limiting naval armaments.

BB-12 • USS OHIO

(BB-12: dp. 12,723; l. 393'10"; b. 72'3"; dr. 23'10"; s. 18 k.; cpl. 561; a. 4 12", 16 6", 6 3", 8 3-pdr., 6 1-pdr., 2 .30 cal. mg.; cl. Maine)

The third ship named Ohio (BB-12) was laid down on April 22, 1899, by Union Iron Works in San Francisco, California. She was launched on May 18, 1901, sponsored by Miss Helen Deshler, and commissioned on October 4, 1904, with Captain Leavitt C. Logan in command.

Designated the flagship of the Asiatic Fleet, Ohio departed San Francisco on April 1, 1905, for Manila. There, she embarked the party of then-Secretary of War William Howard Taft, which included Miss Alice Roosevelt, the President's daughter. She conducted this party on much of its Far Eastern tour of inspection and continued the cruise in Japanese, Chinese, and Philippine waters until returning to the United States in 1907.

Ohio sailed out of Hampton Roads, Virginia, on December 16, 1907, with the battleships of the Atlantic Fleet. The fleet saluted President Theodore Roosevelt in a review that marked the start of its cruise around the world, signifying the emergence of the United States as a major world power.

Commanded by Rear Admiral Robley D. Evans and later Rear Admiral Charles S. Sperry, the fleet visited ports on the east and west coasts of South America, rounding Cape Horn en route to San Francisco. On July 7, 1908, Ohio and her sister ships set course westward to Hawaii, New Zealand, and Australia. Each port welcomed the American ships enthusiastically, but Tokyo, where they anchored on October 18, greeted them with exceptional friendliness. The fleet's presence in Japan symbolized both American friendship and strength, helping to ease strained relations between the two countries.

The fleet visited Amoy, returned to Yokohama, conducted target practice in the Philippines, and began its homeward journey on December 1. After passing through the Suez Canal on January 4, 1909, the fleet made Mediterranean stops before anchoring in Hampton Roads on February 22.

Ohio sailed on to New York, her home port for the next four years, training men of the New York Naval Militia and performing general service with the Atlantic Fleet.

In 1914, she sailed to the Gulf of Mexico to join the patrol off Vera Cruz, safeguarding American interests amidst Mexican political turmoil. Ohio returned north in the summer for a Naval Academy midshipmen cruise, then joined the Reserve Fleet at Philadelphia, recommissioning for each of the next two summers' midshipmen cruises in 1915 and 1916.

After the United States entered World War I, Ohio was recommissioned on April 24, 1917. Throughout the war, she operated out of Norfolk, training crews for the expanding fleet and participating in battleship maneuvers. She arrived in Philadelphia on November 28, 1918, was placed in reserve there on January 7, 1919, decommissioned on May 31, 1922, and sold for scrapping on March 24, 1923.

BB-13 • USS VIRGINIA

(BB-13: dp. 14,980 (tl.); l. 441'3"; b. 76'2~"; dr. 23'9" (mean), s. 19.01 k. (tl.), cpl. 916; a. 4 12", 8 8", 12 6", 12 3", 24 1-pdrs., 4 .30-cal. Colt mg.; 4 21" tt.; cl. Virginia)

The fourth ship with the name of Virginia (Battleship No. 13) was laid down on 21 May 1902 at Newport News, Va., by the Newport News Shipbuilding and Dry Dock Co., launched on 5 April 1904; sponsored by Miss Gay Montague, daughter of the Governor of Virginia; and commissioned on 7 May 1906, Capt. Seaton Schroeder in command.

After fitting out, Virginia conducted her "shaking down" cruise in Lynnhaven Bay, Va., off Newport, R.I., and off Long Island, N.Y., before she put into Bradford, R.I., for coal on 9 August. After running trials for the standardization of her screws off Rockland, Maine, the battleship maneuvered in Long Island Sound before anchoring off President Theodore Roosevelt's home,

Oyster Bay, Long Island, from 2 to 4 September, for a Presidential review.

Virginia then continued her shakedown cruise before she coaled again at Bradford. Meanwhile, events were occurring in the Caribbean that would alter the new battleship's employment. On the island of Cuba, in August of 1906, a revolution had broken out against the government of President T. Estrada Palma. The disaffection, which had started in Pinar del Rio province, grew in the early autumn to the point where President Palma had no recourse but to appeal to the United States for intervention.

By mid-September, it had become apparent that the small Cuban constabulary (3,000 rural guards) was unable to protect foreign interests, and intervention would be necessary. Accordingly, Virginia departed Newport on 15 September 1906, bound for Cuba, and reached Havana on the 21st, ready to disembark General Frederick Funston at Norfolk upon her arrival there and coaled before heading north to Tompkinsville to await further orders. She shifted soon thereafter to the New York Navy Yard where she was coaled and drydocked to have her hull bottom painted before undergoing repairs and alterations at the Norfolk Navy Yard from 3 November 1906 to 18 February 1907. After installation of fire control apparatus at the New York Navy Yard between 19 February and 23 March, the battleship sailed once more for Cuban waters, joining the fleet at Guantanamo Bay on 28 March.

Virginia fired target practices in Cuban waters before she sailed for Hampton Roads on 10 April to participate in the Jamestown Tricentennial Exposition festivities. She remained in Hampton Roads for a month, from 15 April to 15 May, before she underwent repairs at the Norfolk Navy Yard into early June. Subsequently reviewed in Hampton Roads by President Theodore Roosevelt between 7 and 13 June, Virginia shifted northward for target practices on the target grounds of Cape Cod Bay—evolutions that lasted from mid-June to mid-July. She later cruised with her division to Newport; the North River, New York City; and to Provincetown, Mass., before conducting day and night battle practice in Cape Cod Bay.

Returning southward early that autumn, Virginia underwent two months of repairs and alterations at the Norfolk Navy Yard, from 24 September to 24 November, before undergoing further repairs at the New York Navy Yard later in November. She subsequently shifted southward again, reaching Hampton Roads on 6 December.

Virginia spent the next 10 days preparing for a feat never before attempted—a round-the-world cruise by the battleships of the Atlantic Fleet. The voyage, regarded by President Roosevelt as a dramatic gesture to the Japanese—who had only recently emerged on the world stage as a power to be reckoned with—proved to be a signal success, with the ships performing so well as to confound the doomsayers who had predicted a fiasco.

The cruise began eight days before Christmas of 1907 and ended on Washington's Birthday, 22 February 1909. During the course of the voyage, the ships called at ports along both coasts of South America; on the west coast of the United States; at Hawaii, in the Philippines; Japan; China; and in Ceylon. Virginia's division also visited Smyrna, Turkey, via Beirut, during the Mediterranean leg of the cruise. Both upon departure and upon arrival, the fleet was reviewed at Hampton Roads by President Roosevelt, whose "big stick" diplomacy and flair for the dramatic gesture had been practically personified by the cruise of the "Great White Fleet."

Following that momentous circumnavigation, Virginia underwent four months of voyage repairs and alterations at the Norfolk Navy Yard from 26 February to 26 June 1909. She spent the next year and three months operating off the eastern seaboard of the United States, ranging from the southern drill grounds, off the Virginia capes, to Newport, R.I. During that time, she conducted one brief cruise with members of the Naval Militia embarked and visited Rockport and Provincetown, Mass. For the better part of that time, she conducted battle practices with the fleet—evolutions only broken by brief periods of yard work at Norfolk and Boston.

Virginia visited Brest, France, and Gravesend, England, from 16 November to 7 December and from 8 to 29 December 1909, respectively, before she—as part of the 4th Division, Atlantic Fleet—joined the Atlantic Fleet in Guantanamo Bay for drills and exercises. She subsequently operated in Cuban waters for two months, from 13 January to 13 March 1910 before she returned north for battle practices on the southern drill grounds.

Virginia departed Hampton Roads on 11 April, in company with Georgia (Battleship No. 15), and reached the Boston Navy Yard two days later. She underwent repairs there until 24 May before putting to sea for Provincetown. Over the next five days, Virginia operated with the collier Vestal, testing a "coaling-at-sea apparatus" off Provincetown and at Stellwagen's Bank before she conducted torpedo practices. The battleship returned to the Boston Navy Yard on 18 June.

Virginia maintained her routine of operations off the eastern seaboard—occasionally ranging into Cuban waters for regularly scheduled fleet evolutions in tactics and gunnery—into 1913, a routine largely uninterrupted. In 1913, however, unrest in Mexico caused the frequent dispatch of American men-of-war to those waters. Virginia became one of those ships in mid-February, when she reached Tampico on the 15th of that month; she remained there until 2 March, when she shifted to Vera Cruz for coal. She returned to Tampico on 5 March and remained there for 10 days.

After another stint of operations off the eastern seaboard, ranging from the Virginia capes to Newport—a period of maneuvers and exercises varied by a visit to New York at the end of May 1913 for the dedication of the memorial to the battleship Maine (sunk in Havana Harbor in February 1898) and one to Boston in mid-June for Flag Day and Bunker Hill exercises—Virginia returned to Mexican waters in November. She reached Vera Cruz on 4 November and remained in port until the 30th, when she shifted to Tampico. She observed conditions in those ports and operated off the Mexican coast into January of 1914.

Returning to Cuban waters for exercises and maneuvers with the fleet, Virginia sailed for the Virginia capes in mid-March 1914. She maneuvered with the fleet off Cape Henry and in Lynnhaven Roads before she conducted gunnery drills at the wreck of San Marcos (ex-Texas) in Tangier Sound, Chesapeake Bay. Virginia subsequently held experimental gunnery firings on the southern drill grounds before she spent much of April drydocked at Boston.

The American occupation of Vera Cruz in April 1914 resulted in the sizable deployment of American men-of-war to that port that lasted into the autumn. Virginia reached Vera Cruz on 1 May and operated with the fleet out of that port into early October, a period of time broken by target practice in Guantanamo Bay between 18 September and 3 October.

While war raged in Europe, Virginia continued her operations off the eastern seaboard of the United States, ranging from the southern drill grounds to the coast of New England and occasionally steaming to Cuban waters for winter maneuvers. She was placed in reserve on 20 March 1916, at the Boston Navy Yard, and was undergoing an extensive overhaul in the spring of 1917 when the United States declared war on Germany.

On the day America entered World War I, the United States government took steps to take over all interned German merchant vessels then in American ports. As part of that move, Virginia sent boarding parties to seize the German passenger and cargo vessels America, Cincinnati, Wittekind, Koln, and Ockenfels on 6 April 1917.

Completing her overhaul at Boston on 27 August, Virginia sailed for Port Jefferson, N.Y., three days later, to join the 3rd Division, Battleship Force, Atlantic Fleet. Over the ensuing 12 months, the battleship served as a gunnery training ship out of Port Jefferson and Norfolk, with service interrupted briefly in early December 1917 when she became the temporary flagship for Rear Admiral John A. Hoogewerff, Commander, Battleship Division 1. She subsequently became the flagship for the 3rd Division commander, Rear Admiral Thomas Snowden.

Overhauled at the Boston Navy Yard in the autumn of 1918, Virginia spent the remainder of hostilities engaged in convoy escort duties, escorting convoys well over halfway across the Atlantic. She departed New York on 14 October 1918 on her first such mission, covering a convoy that had some 12,176 men embarked. After escorting those ships to longitude 22 degrees west, she put about and headed for home.

That proved to be her only such wartime mission; however, the armistice was signed on 11 November 1918, the day before Virginia set out with a France-bound convoy, her second escort run into the mid-Atlantic. After leaving that convoy at longitude 34 degrees west, Virginia put about and headed for Hampton Roads.

The cessation of hostilities meant the return of the many troops who had been engaged in fighting the enemy overseas. Similar in mission to the 'Magic Carpet' operation that followed the end of World War II, a massive troop-lift, bringing the 'doughboys' back from 'over there,' commenced soon after World War I ended.

With additional messing and berthing facilities installed to permit her use as a troopship, Virginia departed Norfolk eight days before Christmas of 1918. Over the ensuing months, she conducted five round-trip voyages to Brest, France, and back. Reaching Boston on Independence Day 1919, ending her last troop lift, Virginia concluded her transport service, having brought some 6,037 men back from France.

Virginia remained at the Boston Navy Yard, inactive until decommissioned there on 13 August 1920. Struck from the Navy list and placed on the sale list on 12 July 1922, the battleship, reclassified prior to her inactivation to BB-13 on 17 July 1920, was subsequently taken off the sale list and transferred to the War Department on 6 August 1923 for use as a bombing target.

Virginia and her sister ship New Jersey were taken to a point three miles off the Diamond Shoals lightship, off Cape Hatteras, N.C., and anchored there on 5 September 1923. The 'attacks' made by Army Air Service Martin bombers began shortly before 0900. On the third attack, seven Martins, flying at 3,000 feet, each dropped two 1,100-pound bombs on Virginia; only one of them hit.

That single bomb, however, 'completely demolished the ship as such.' An observer later wrote: 'Both masts, the bridge, all three smokestacks, and the upperworks disappeared with the explosion, and there remained, after the smoke cleared away, nothing but the bare hull, decks blown off, and covered with a mass of tangled debris from stem to stern consisting of stacks, ventilators, cage masts, and bridges.'

Within one-half hour of the cataclysmic blast that wrecked the ship, her battered hulk sank beneath the waves. Her sister ship ultimately joined her shortly thereafter. Virginia's end and New Jersey's provided far-sighted naval officers with a dramatic demonstration of air power and impressed upon them the 'urgent need of developing naval aviation with the fleet.' As such, the service performed by the old pre-dreadnought may have been her most valuable.

BB-14 • USS NEBRASKA

(BB-14: dp. 16,094; l. 441'3"; b. 76'2"; dr. 25'10"; s. 19 k.; cpl. 1,108; a. 4 12", 8 8", 12 6", 12 3", 4 21" tt.; cl. Virginia)

Nebraska (BB-14), ex-Pennsylvania, was laid down by Moran Brothers in Seattle, Washington, on 4 July 1902; launched on 7 October 1904; sponsored by Miss Mary N. Mickey, daughter of Governor John H. Mickey of Nebraska, and commissioned on 1 July 1907, Captain Reginald F. Nicholson in command.

After shakedown and alterations, the new battleship joined the "Great White Fleet" at San Francisco after 6 May 1908, replacing Alabama (BB-8).

Departing San Francisco on 7 July 1908, the Fleet visited Honolulu, Hawaii; Auckland, New Zealand; Sydney and Melbourne, Australia; Manila, Philippine Islands; Yokohama, Japan; and Colombo, Ceylon, arriving in Suez, Egypt, on 3 January 1909. Departing Messina, Italy, on the 9th, the Fleet visited Naples, Italy, and then Gibraltar, arriving at Hampton Roads on 22

February where President Theodore Roosevelt reviewed the Fleet as it passed into the roadstead.

Nebraska continued duty with the Atlantic Fleet. She attended the Hudson-Fulton Celebration in 1910 and the Louisiana Centennial in 1912. She earned the Mexican Service Medal for operations at Vera Cruz, Mexico, from 1 May to 21 June 1914 and from 1 June to 13 October 1916. After a period of reduced commissioned service, she was again placed in full commission on 3 April 1917.

When war was declared on 6 April 1917, Nebraska was undergoing repairs at the Boston Navy Yard, attached to the 3rd Division, Battleship Force, U.S. Atlantic Fleet. On 13 April 1917, she departed Boston to engage in maneuvers and battle practice with the fleet in the Chesapeake Bay area. She operated along the East Coast, primarily training armed guard crews for American merchantmen, until entering the Norfolk Navy Yard on 15 April 1918 for repairs.

At Hampton Roads on 16 May, she received on board the body of the late Carlos M. DePena, Envoy Extraordinary and Minister Plenipotentiary from Uruguay, with full honors, departing Hampton Roads the same day and arriving in Montevideo on 10 June in company with Pittsburgh (ACR-4), flagship of the Pacific Fleet. The Commander in Chief, U.S. Pacific Fleet, came on board for the ceremonies and the body of the late Uruguayan Minister to the United States was transferred with full honors. Nebraska departed Montevideo on 15 June for home, arriving at Hampton Roads on 26 July.

The battleship departed New York on 17 September as principal escort for a fast merchant convoy of 18 ships to an eastern Atlantic rendezvous, returning to Hampton Roads on 3 October. Nebraska made two more convoy voyages in the Atlantic, returning from the latter on 2 December to prepare for service in returning American troops from France.

Nebraska made four voyages from the United States to Brest, France, transporting 4,540 troops to and from the United States. On her first trip, she departed Hampton Roads on 30 December 1918, arrived in Brest on 11 January 1919, and returned to Newport News on 28 January. The final voyage to return veterans from France ended when she arrived in Newport News, Virginia, on 21 June with 1,279 troops.

On 22 June 1919, Nebraska was detached from the transport service and shortly thereafter sailed to join Division 2, Squadron 1, U.S. Pacific Fleet, for operations along the West Coast under the command of Captain P. N. Olmstead until she decommissioned on 2 July 1920.

In accordance with the Washington Treaty limiting naval armament, Nebraska was rendered incapable of further warlike service on 9 November 1923 and sold for scrap a few weeks later.

BB-15 • USS GEORGIA

(BB-15: dp. 16,094; l. 441'3"; b. 76'2"; dr. 25'10"; s. 19 k.; cpl. 1,108; a. 4 12", 8 8", 12 6", 12 3", 4 21" tt.; cl. Virginia)

Georgia was launched by the Bath Iron Works of Bath, Maine, 11 October 1904, sponsored by Miss Stella Tate, and commissioned at Boston Navy Yard 24 September 1906, Captain R. G. Davenport in command.

After Georgia was fitted out and completed a short shakedown cruise, she Joined the Atlantic Fleet as flagship of Division 2, Squadron 1. Georgia departed Hampton Roads 26 March 1907 for Guantanamo Bay, Cuba, where she participated in gunnery practice with the fleet. After returning briefly to Boston Navy Yard for repairs, Georgia joined with other ships of the Atlantic Fleet in ceremonies opening the Jamestown Exposition. President Roosevelt and dignitaries present reviewed the fleet 10 June 1907, and 11 June was proclaimed "Georgia Day" at the exposition in special ceremonies aboard Georgia.

Georgia next sailed with the fleet for target practice in Cape Cod Bay, arriving 15 June. During these drills 15 July, a powder charge ignited prematurely in her aft 8" turret, killing 10 officers and men and injuring 11. Condolences for the loss from this tragic accident were received from all over the world.

The powerful battleship then participated in the tercentennary of the landing of the first English Colonists 16 to 21 August 1907, after which she rejoined the fleet for battle maneuvers before mooring at League Island, N.Y., 24 September, for overhaul.

Arriving in Hampton Roads 7 December 1907, Georgia gathered with 15 other battleships, a torpedo boat squadron, and transports for the great naval review preceding the cruise of the Atlantic Fleet to the West Coast. On 16 December President Roosevelt reviewed the assembled "Great White Fleet" and sent it on the first leg of an around-the-world voyage of training, and building of American prestige and good will. Visiting many South American countries on their highly successful cruise, the fleet met with ships of the Pacific Fleet in another review in San Francisco Bay for the Secretary of the Navy 8 May 1908. Then Georgia, in company with other battleships and supply vessels, departed San Francisco 7 July 1908 for the second leg of the cruise, showing the flag and bringing the message of American sea power to many parts of the world, including the Philippine Islands, Australia, Japan, and Mediterranean ports. The fleet returned to Hampton Roads 22 February 1909.

Georgia continued to serve with the Atlantic Fleet in exercises and battle maneuvers, with periods of overhaul interspersed, until 2 November 1910 when President Taft reviewed the fleet prior to its departure for France. In an elaborate battle and scouting problem, Georgia and the other battleships continued their training, visiting Weymouth, England, and returning to Guantanamo Bay, Cuba, 13 March 1911.

From 1911 to 1913, Georgia continued to train and serve as a ceremonial ship and 5 June 1913 participated in a 2-month practice cruise for Naval Academy Midshipmen. After a long overhaul period in Boston Navy Yard, Georgia arrived off the coast of Mexico 14 January 1914 with other fleet units to protect American interests in the troubled Vera Cruz-Tampico area. The busy battleship returned briefly to Norfolk, VA., in March, but was soon back cruising Mexican waters, and from August to October 1914 cruised off Haiti for the protection of American civilians in that country.

After another period of overhaul, Georgia joined the fleet off Cuba 25 February 1915 for winter maneuvers, and spent the rest of the year in training and ceremonial duties with the Atlantic Fleet Battleship Force. She arrived at Boston Navy Yard for overhaul 20 December 1915 and decommissioned 27 January 1916.

Assigned as a receiving ship at Boston, Georgia was called to duty at the outbreak of World War I, and commissioned again 6 April 1917. For the next 18 months, she operated with the 3d Division, Battleship Force, in fleet tactical exercises and merchant crew gunnery training, tensed in the York River, VA She joined with Cruiser Force Atlantic briefly in September 1918 to escort convoys to meet their eastern escorts, and beginning 10 December 1918 was fitted out as a transport and attached to the Cruiser and Transport Force for the purpose of returning troops of the A.E.F. to the United States. Georgia made five voyages to France from December 1918 to June 1919 and brought home nearly 6,000 soldiers.

Georgia was next transferred to the Pacific Fleet as flagship of Division 2, Squadron 1. She left Boston for San Diego, via the Panama Canal, 16 July 1919, and after participating in ceremonial operations for 2 months. entered Mare Island Naval Shipyard for repairs 20 September 1919. Here Georgia stayed Until decommissioning 10 July 1920. She was eventually sold for scrap 1 November 1923 in accordance with the Washington Treaty for the limitation of naval armaments, and her name was struck from the Navy List 10 November 1923.

BB-16 • USS NEW JERSEY

(BB-16: dp. 14,948; l. 441'3"; b. 76'3"; dr. 23'9"; s. 19 k.; cpl. 812; a. 4 12", 8 8", 12 6", 4 21" tt.; cl. Virginia)

The first New Jersey (BB-16) was launched on 10 November 1904 by the Fore River Shipbuilding Company, Quincy, Mass., sponsored by Mrs. William B. Kenney, daughter of Governor Franklin B. Murphy of New Jersey; and commissioned on 12 May 1906, Captain William W. Kimball in command.

New Jersey's initial training in Atlantic and Caribbean waters was highlighted by her review by President Theodore Roosevelt in Oyster Bay during September 1906, and by her presence in Havana, Cuba, from 21 September through 13 October to protect American lives and property threatened by the Cuban Insurrection. From 15 April to 14 May 1907, she lay in Hampton Roads representing the Navy at the Jamestown Exposition.

In company with fifteen other battleships and six attendant destroyers, New Jersey cleared Hampton

Roads on 16 December 1907, her rails manned and her guns firing a 21-gun salute to President Roosevelt, who watched from Mayflower the beginning of the dramatic cruise of the Great White Fleet. The international situation required a compelling exhibition of the strength of the United States; this round-the-world cruise was to provide one of the most remarkable illustrations of the ability of sea power to keep peace without warlike action. Not only was a threatened conflict with Japan averted, but notice was served on the world that the United States had come of age and was an international power that could make its influence felt in any part of the world.

Commanded first by Rear Admiral Robley D. Evans, and later by Rear Admiral Charles S. Sperry, the fleet laid its course for Trinidad and Rio de Janeiro, then rounded Cape Horn. After calling at Punta Arenas, Valparaiso, and Callao, the battleships made a triumphant return to the United States at San Francisco. On 7 July 1908, the fleet sailed west, bound for Hawaii, Auckland, and three

Australian ports; Sydney, Melbourne, and Albany. Each city seemed to offer a more enthusiastic reception for the American sailors and their powerful ships than the last, but tension and rumor of possible incident made the arrival in Tokyo Bay on 18 October unique among the cruise's calls.

Immediately it was clear that no special precautions had been necessary; nowhere during the cruise did the men of New Jersey and her sisters meet with more expressions of friendship, both through elaborately planned entertainment and spontaneous demonstrations. The President observed with satisfaction this accomplishment of his greatest hope for the cruise: "The most noteworthy incident of the cruise was the reception given to our fleet in Japan."

The Great White Fleet sailed on to Amoy, returned briefly to Yokohama, then held target practice in the Philippines before beginning the long homeward passage on 1 December. The battleships passed through the Suez Canal on 4 January 1909, called at Port Said, Naples, and Villefranche, and left Gibraltar astern on 6 February. In one of the last ceremonial acts of his presidency, Theodore Roosevelt reviewed the Great White Fleet as it anchored in Hampton Roads on 22 February.

Aside from a period out of commission in reserve at Boston from 2 May 1910 until 15 July 1911, New Jersey carried out a normal pattern of drills and training in the Western Atlantic and Caribbean, carrying

midshipmen of the United States Naval Academy in the summers of 1912 and 1913. With Mexican political turmoil threatening American interests, New Jersey was ordered to the Western Caribbean in the fall of 1913 to provide protection. On 21 April 1914, as part of the force commanded by Rear Admiral Frank F. Fletcher, following the Mexican refusal to apologize for an insult to American naval forces at Tampico, sailors and marines landed at Vera Cruz and took possession of the city and its customs house until changes in the Mexican government made evacuation possible. New Jersey sailed from Vera Cruz on 13 August, observed and reported on troubled conditions in Santo Domingo and Haiti, and reached Hampton Roads on 9 October. Until the outbreak of World War I, she returned to her regular operations along the East Coast and in the Caribbean.

During World War I, New Jersey made a major contribution to the expansion of the wartime Navy, training gunners and seamen recruits in Chesapeake Bay. After the Armistice, she began the first of four voyages to France, from which she had brought home 5,000 members of the AEF by 9 June 1919. New Jersey was decommissioned at the Boston Naval Shipyard on 6 August 1920 and was sunk off Cape Hatteras on 5 September 1923 in Army bomb tests conducted by Brig. Gen. William Mitchell.

BB-17 • USS RHODE ISLAND

(BB-17: dp. 14,948 (n.),l. 441'8" b. 76'3", dr. 23'9", s. 19 k.; cpl. 812; a. 4 12", 8 8", 12 6", 12 3", 12 3-pdr., 4 21" tt.; cl. Virginia)

The second Rhode Island was launched 17 May 1904 by Fore River Shipbuilding Co., Quincy, Mass.; sponsored by Mrs. F. C. Dumaine; and commissioned 19 February 1906 Capt. Perry Garst in command.

Rhode Island underwent extensive shakedown and acceptance trials on the U.S. east coast between Hampton Roads and Boston before being assigned to Division 2, Squadron 1, Atlantic Fleet 1 January 1907. The battleship departed Hampton Roads 9 March 1907 for Guantanamo Bay, Cuba, to participate in gunnery practice and squadron operations evolutions.

She then returned north to cruise between Hampton Roads and Cape Cod Bay. Arriving in Hampton Roads 8 December 1907, Rhode Island joined 15 other battleships, a torpedo boat squadron, and transports, for the great fleet review which began the cruise of the Atlantic Fleet to the west coast and around the w orld. President Theodore Roosevelt reviewed the "Great White Fleet" 16 December and sent it on the first leg of the long voyage. Rhode Island called at Trinidad, British West Indies Rio de Janeiro, Punta Arenas, Callao, and Magdalena Bay before arriving at San Diego, Calif., 14 April 1908.

The fleet remained on the west coast into July, Rhode Island steaming north to visit the Puget Sound area during June. The entire fleet departed San Francisco 7 July 1908 for Honolulu, Auckland, Sydney, Melbourne, and Manila, arriving in the Philippines 2 October. From Manila Rhode Island made for Yokohama, Japan, returning to Olongapo, Philippine Islands, at the end of October. Departing Cavite 1 December Rhode Island visited Colombo, Suez, Marseille, and Gibraltar before returning to Hampton Roads 22 February 1909.

Subsequently entering New York Navy Yard for overhaul, Rhode Island was reassigned 8 March 1909 to Division 3 Squadron 1. She continued to serve with the Atlantic Fleet into 1910 participating in exercises including deployment southward to the Caribbean during February 1910. Assigned 20 October 1910 to Division 4, Squadron 1, Rhode Island and other fleet units were reviewed 2 November at Boston by President Taft prior to their departure for European waters. In an elaborate battle and scouting problem, the fleet continued its training, Rhode Island subsequently visiting Gravesend, England, before returning to Guantanamo Bay 13 January 1911.

Rhode Island continued her duties attached to the Atlantic Fleet up to the outbreak of war in Europe in 1914. She cruised southward to Key West, Havana, and Guantanamo Bay during June and July 1912 but otherwise remained on the east coast operating between Hampton Roads and Rockland Maine. Reassigned to Division 3, Squadron 1, Atlantic Fleet Rhode Island became division flagship 17 July 1912. She transferred the division flag to New Jersey 1 August in the periodic rotation of additional flag duties among units of her division.

The Commander, Division 3, Squadron 1 transferred his flag from Virginia to Rhode Island 28 June 1913 and remained on board until 18 January 1914. At the end of 1913, Rhode Island cruised off the Mexican coast to protect citizens and property threatened by deteriorating political developments ashore. Arriving off Vera Cruz 4 November 1913, Rhode Island operated off Tampico and Tuxpan into February 1914. After 2 weeks at Guantanamo Bay the battleship returned northward to Virginia waters.

Rhode Island kept up her continuous schedule of annual docking and overhaul gunnery practice, and squadron maneuvers well into 1916. She remained off the U.S. eastern seaboard but occasionally steamed into more southerly waters she called at Caribbean ports during October 1914 to March 1915 and January to February 1916. Rhode Island undertook additional duty as flagship, Division 4, Squadron 1, from 19 December 1914 until 20 January 1915.

Placed in reduced commission in reserve 15 May 1916 at Boston Navy Yard, Rhode Island was detached from the Atlantic Fleet the following day. The battleship flew the flag of the Commander-in-Chief, Reserve Force, Atlantic Fleet from 24 June 1916 to 28 September.

Returned to full commission 27 March 1917 at Hampton Roads, Rhode Island broke the flag of the Commander, Battleship Division .3, Atlantic Fleet, 3 May 1917 shortly after the United States entered World War I. Undertaking vigorous gunnery practice and emergency drills to reach combat readiness, Rhode Island was assigned antisubmarine patrol duty off Tangier Island, Md. Based at Hampton Roads into 1918 Rhode Island was transferred to Battleship Division 2 during April. Remaining ready for overseas deployment, Rhode Island undertook special torpedo proving trials during June 1918.

Upon the war's end in November 1918, Rhode Island was ordered to assist returning U.S. troops from France. Fitted with hundreds of extra bunks, the battleship made five roundtrip voyages across the Atlantic between 18 December 1918 and 4 July 1919. In all she transported over 5,000 men from Brest, France, to Hampton Roads and Boston.

Designated flagship of Battleship Squadron 1, Pacific Fleet 17 July 1919 at Boston, Rhode Island departed Boston Navy Yard 24 July for Balboa, C.Z., and Mare Island Navy Yard to undertake her new assignment. After remaining at Mare Island into 1920, Rhode Island decommissioned 30 June and was placed in reserve.

Rendered incapable of any further warlike service 4 October 1923 in accordance with the Washington Treaty limiting naval armaments, Rhode Island was sold 1 November 1923 for Scrapping.

BB-18 • USS CONNECTICUT

(BB-18: dp. 16, 000; l. 456'4"; b. 76'10"; dr. 24'6"; s. 18 k.; cpl. 827; a. 4 12", 8 8", 12 7"; cl. Connecticut)

The fourth ship named Connecticut (BB-18) was launched on 29 September 1904 by the New York Navy Yard, sponsored by Miss A. Welles, granddaughter of Gideon Welles, Secretary of the Navy during the Civil War, and commissioned on 29 September 1906, with Captain W. Swift in command.

The Connecticut-class battleships were a leap forward in naval design for the U.S. Navy. Connecticut, with a displacement of 16,000 tons, was larger and more heavily armed than her predecessors. Her main battery consisted of four 12-inch guns, complemented by an array of eight 8-inch guns and twelve 7-inch guns.

After joining the Atlantic Fleet, Connecticut became its flagship on 16 April 1907. Later that month, she participated in the Presidential Fleet Review and other

ceremonies marking the opening of the Jamestown Exposition. On 16 December 1907, still as the flagship, she embarked from Hampton Roads on the Great White Fleet's world cruise. On 8 May 1908, the Atlantic Fleet joined the Pacific Fleet in San Francisco Bay for a review by the Secretary of the Navy, and the combined fleets, with Connecticut as the flagship, continued their cruise, showcasing American strength around the world. The fleet returned to Hampton Roads on 22 February 1909.

Connecticut continued as the flagship of the Atlantic Fleet until 1912, cruising the East Coast and the Caribbean from its base at Norfolk, conducting training, and participating in ceremonial events. From 2 November 1910 to 17 March 1911, she embarked on an extended scouting problem in European waters. Between 1913 and 1915, Connecticut served with the Fourth Division of the Atlantic Fleet, usually as the flagship. Apart from a brief Mediterranean cruise in October and November 1913, she operated in the Caribbean, protecting American citizens and interests during unrest in Mexico and Haiti.

After repairs and a stint as a receiving ship at the Philadelphia Navy Yard in 1916, Connecticut returned to full commission on 3 October 1916 as the flagship of the Fifth Division, Battleship Force, Atlantic Fleet. She operated along the East Coast and in the Caribbean until the United States entered World War I. During the war, based in the York River, Virginia, she trained in Chesapeake Bay, preparing midshipmen and gun crews for merchant ships. At the war's end, she was outfitted for transport duty and between 6 January and 22 June 1919, made four voyages to bring troops back from France. On 23 June 1919, she was reassigned as the flagship of Battleship Squadron 2, Atlantic Fleet.

In the summer of 1920, Connecticut sailed to the Caribbean and the West Coast on a midshipman-Naval Reserve training cruise. The next summer, she visited European ports on similar duty, and upon her return to Philadelphia on 21 August 1921, was assigned as the flagship of the Train, Pacific Fleet. She arrived at San Pedro, California, on 28 October, and over the next year, cruised along the West Coast, participating in exercises and commemorations. Entering Puget Sound Navy Yard on 16 December 1922, Connecticut was decommissioned there on 1 March 1923 and sold for scrapping on 1 November 1923, in accordance with the Washington Treaty for the limitation of naval armaments.

BB-19 • USS LOUISIANA

(BB-19: Displacement: 16,000; Length: 456'4"; Beam: 76'10"; Draft: 24'6"; Speed: 18 knots; Complement: 827; Armament: 4 12-inch guns, 8 8-inch guns, 12 7-inch guns, 20 3-inch guns, 12 3-pounders, 2 1-pounders, 4 .30 cal. machine guns, 4 21-inch torpedo tubes; Class: Connecticut)

The third ship named Louisiana (BB-19) was laid down on 7 February 1903 by the Newport News Shipbuilding & Dry Dock Co., Newport News, Virginia. She was launched on 27 August 1904, sponsored by Miss Juanita LaLande, and commissioned on 2 June 1906, with Captain Albert R. Couden in command.

Following her shakedown off the New England coast, Louisiana sailed on 15 September for Havana in response to an appeal by Cuban President Estrada Palma for American assistance in suppressing an insurrection. The new battleship carried a peace commission comprised of Secretary of War William H. Taft and Assistant Secretary of State Robert Bacon, which arranged for a provisional

government on the island. Louisiana stood by while this government was set up and then returned the commission to Fortress Monroe, Virginia.

Louisiana embarked President Theodore Roosevelt at Piney Point, Maryland, on 8 November for a cruise to Panama to inspect work on the construction of the Panama Canal. Returning, she briefly visited Puerto Rico, where the President studied the administrative structure of the Commonwealth's government, before debarking him at Piney Point on 26 November.

During 1906 and 1907, Louisiana visited New Orleans, Havana, and Norfolk; maneuvered out of Guantanamo Bay; and engaged in battle practice along the New England coast. On 16 December 1907, she departed Hampton Roads as one of the 16 battleships President Theodore Roosevelt sent on a voyage around the world. The cruise of the "Great White Fleet" deterred hostile actions towards the United States by other countries, primarily Japan, raised American prestige as a global naval power, and impressed upon Congress the importance of a strong Navy and a thriving merchant fleet. During the circumnavigation, Louisiana visited Port-of-Spain; Rio de Janeiro; Punta Arenas and Valparaiso, Chile; Callao, Peru; San Diego and San Francisco; Honolulu; Auckland; Sydney; Tokyo; Manila; Amoy, China; Hong Kong; Manila; Colombo; Suez and Port Said; Smyrna; and Gibraltar before returning home on 22 February 1909.

After overhaul and maneuvers, Louisiana joined the 2nd Division of the Atlantic Fleet on 1 November 1910 and sailed for European waters to visit English and French ports before returning to the United States in the spring of 1911. During the summer, she paid formal visits to the north European ports of Copenhagen; Trollhättan, Sweden; Kronstadt, Finland; and Kiel, Germany, and was inspected by the Kings of Denmark and Sweden, the Kaiser, and the Tsar.

Between 6 July 1913 and 24 September 1915, Louisiana made three voyages from east coast ports to Mexican waters. On the first (6 July to 23 December 1913), she stood by to protect American lives and property and to help enforce both the Monroe Doctrine and the arms embargo which had been established to discourage further revolutionary disturbances in Mexico. Her second voyage (14 April to 8 August 1914) came at a time when tension between Mexico and the United States was at its peak during the shelling and occupation of Vera Cruz. Louisiana sailed a third time for Mexican waters to protect American interests again from 17 August to 24 September 1915.

Returning from the Gulf of Mexico, Louisiana was placed in reserve at Norfolk and, until the United States entered World War I, she served as a training ship for midshipmen and naval militiamen on summer cruises.

During World War I, Louisiana was assigned as a gunnery and engineering training ship, cruising off the middle Atlantic coast until 25 September 1918. At that time, she became one of the escorts for a convoy to Halifax. Beginning 24 December, she saw duty as a troop transport, making four voyages to Brest, France, to carry troops back to the United States.

Following her final trip back from Brest, Louisiana reported to the Philadelphia Navy Yard, where she decommissioned on 20 October 1920 and was sold for scrap on 1 November 1923.

BB-20 • USS VERMONT

(BB-20: dp. 16,000 (n.); l. 456'4"; b. 76'10"; dr. 24'6" (mean), s. 18 k., cpl. 880, a. 4 12", 8 8", 12 7", 20 3", 12 3-pdrs., 4 1-pdrs, 4 .30-cal. mg., 2 .30-cal. Colt mg.; cl. Connecticut)

The second Vermont (Battleship No. 20) was laid down on 21 May 1904 at Quincy, Mass., by the Fore River Shipbuilding Co., launched on 31 August 1905, sponsored by Miss Jennie Bell, the daughter of Governor Charles J. Bell of Vermont, and commissioned at the Boston Navy Yard on 4 March 1907, Capt. William P. Potter in command.

After her "shakedown" cruise off the eastern seaboard between Boston and Hampton Roads, Va., Vermont participated in maneuvers with the 1st Division of the Atlantic Fleet and, later, with the 1st and 2d Squadrons. Making a final trial trip between Hampton Roads and Provincetown, Mass., between 30 August and 5 September, Vermont arrived at the Boston Navy

Yard on 7 September and underwent repairs until late in November 1907.

Departing Boston on 30 November, she coaled at Bradford, R.I., received "mine outfits and stores" at Newport, R.I., and picked up ammunition at Tompkinsville, Staten Island, N.Y.; and arrived at Hampton Roads on 8 December.

There, she made final preparations for the globe-girdling cruise of the United States Atlantic Fleet. Nicknamed the "Great White Fleet" because of the white and spar color of their paint schemes, the 16 pre-dreadnought battleships sailed from Hampton Roads on 16 December, standing out to sea under the gaze of President Theodore Roosevelt who had dispatched the ships around the globe as a dramatic gesture toward Japan, a growing power on the world stage.

Vermont sailed as a unit of the 1st Division, under the overall command of Rear Admiral Robley D.

"Fighting Bob" Evans, who was concurrently the Commander in Chief of the Fleet. Over the ensuing months, the battleship visited ports in Chile, Peru, Mexico, California, Hawaii, New Zealand, Australia, the Philippines, Japan, China, and in the Mediterranean, before she returned to Hampton Roads—again passing in review before President Roosevelt—on Washington's Birthday, 22 February 1909. During the voyage, Vermont's commanding officer, Capt. Potter, was advanced to flag rank and took command of the division, his place was taken by Capt. (later Admiral) Frank Friday Fletcher.

Following her return to the United States, Vermont underwent repairs at the Boston Navy Yard from 9 March to 23 June and then rejoined the fleet off Provincetown. She subsequently spent the 4th of July at Boston as part of the 1st Division of the Fleet before spending nearly a month, from 7 July to 4 August, in exercises with the Atlantic Fleet. Subsequently coaling at Hampton Roads, the battleship conducted target practice off the Virginia capes in the operating area known as the Southern Drill Grounds.

For the remainder of 1909, Vermont continued maneuvers and exercises, broken by visits to Stamford, Conn., for Columbus Day festivities and to New York City for the observances of the Hudson-Fulton Celebration from 22 September to 9 October. She spent the Christmas holidays at New York City, anchored in the North River.

The battleship then moved south for the winter, reaching Guantanamo Bay on 12 January 1910. For the next two months, she exercised in those Caribbean climes, returning to Hampton Roads and the Virginia capes for elementary target practice that spring. Ultimately reaching Boston on 29 April, the battleship

underwent repairs at that yard through mid-July, before embarking members of the Naval Militia at Boston for operations between that port and Provincetown from 22 to 31 July.

Vermont subsequently visited Newport and then sailed for Hampton Roads on 22 August, where she then prepared for target practices between 25 and 27 September, before visiting New York City with other ships of the Atlantic Fleet.

After minor repairs at the Philadelphia Navy Yard, the battleship sailed for European waters on 1 November. Reaching the British Isles a little over two weeks later, Vermont—with other units of the 3rd Division, Atlantic Fleet—visited Gravesend, England, from 16 November to 7 December and then called at Brest, France, where she remained until heading for the West Indies on 30 December.

Vermont engaged in winter maneuvers and drills out of Guantanamo Bay, Cuba, from 13 January 1911 to 13 March, before sailing for Hampton Roads. In the ensuing weeks, the battleship operated in the Southern Drill Grounds and off Tangier Island in the Chesapeake Bay, where she conducted target practice. After dropping off target materials at Hampton Roads on 8 April, Vermont sailed later that day for Philadelphia where she arrived on 10 April and entered dry dock.

Later in the spring, Vermont resumed her operations with the other pre-dreadnought battleships of the 3rd Division. She operated off Pensacola, Fla., and ranged into the Gulf of Mexico, calling at Galveston, Tex., from 7 to 12 June before returning to Pensacola on 13 June for provisions.

Shifting northward to Bar Harbor, Maine, Vermont spent the 4th of July there before she drilled and

exercised with the Fleet in Cape Cod Bay and off Provincetown. The battleship then operated off the New England seaboard through mid-August, breaking her periods at sea with a port visit to Salem and alterations at the Boston Navy Yard. She then shifted south to conduct experimental gunnery firings and autumn target practice in the regions from Tangier Sound to the Southern Drill Grounds.

After repairs at the Norfolk Navy Yard from 12 September to 9 October, Vermont rejoined the Fleet at Hampton Roads before participating in the naval review in the North River, at New York City, between 24 October and 2 November. She then maneuvered and exercised with the 1st Squadron of the Fleet before returning to Hampton Roads.

Touching briefly at Tompkinsville on 7 and 8 December, Vermont reached the New York Navy Yard on the latter day for year-end leave and upkeep and remained there until 2 January 1912, when she sailed for the Caribbean and the annual winter maneuvers. She operated in Cuban waters, out of Guantanamo Bay and off Cape Cruz, until 9 March, when she sailed for the Norfolk Navy Yard and an overhaul that lasted into the autumn.

She departed Norfolk on 8 October and reached New York City on the 10th. She participated in the naval review at that port from 10 to 15 October before embarking Commander, 2nd Division, Atlantic Fleet, at Hampton Roads between 16 and 18 October.

Vermont subsequently worked out of Hampton Roads, in the Virginia capes Southern Drill Grounds area, into December. During that time, she conducted target practices and twice participated in humanitarian deeds searching for the stranded steamship SS Noruega on 2 November and assisting the submarine B-2 (Submarine No. 11) between 13 and 15 December.

The battleship spent Christmas 1912 at the Norfolk Navy Yard before steaming for Cuba and winter maneuvers. En route, she visited Colon, Panama, a terminus of the nearly completed Panama Canal,

and reached Guantanamo Bay on 19 January 1913. She subsequently operated out of Guantanamo and Guayancanabo Bay until sailing for Mexican waters on 12 February.

Vermont arrived at Vera Cruz on the 17th and remained at that port into the spring, protecting American interests until 29 April, when she sailed north to rejoin the fleet in Hampton Roads. The battleship conducted one midshipman's training cruise that summer, embarking the midshipmen at Annapolis on 6 June. After rejoining the fleet, Vermont cruised in Block Island Sound and visited Newport.

The battleship then received her regular overhaul at Norfolk from July into October before she conducted target practice off the Southern Drill Grounds. Vermont then made her second European cruise, departing Hampton Roads for French waters on 25 October, reaching Marseilles on 8 November. Ultimately departing that Mediterranean port on 1 December, Vermont reached the Norfolk Navy Yard five days before Christmas, making port on the end of a towline because of storm damage to a propeller.

Soon after she had completed her post-repair trials and had begun preparations for the spring target practice with the Fleet in the Southern Drill Grounds, tension in Mexico beckoned the battleship. Departing Hampton Roads on 15 April, Vermont reached Veracruz very early in the morning of 22 April in company with Arkansas

(Battleship No. 33), New Hampshire (Battleship No. 25), South Carolina (Battleship No. 26), and New Jersey (Battleship No. 16). Her landing force—a "battalion" of 12 officers and 308 men—went ashore after daybreak that same day as United States forces occupied the port to block an arms shipment to the dictator Victoriano Huerta. In the fighting that ensued, two officers from the staff were awarded Medals of Honor: Lt. Julius C. Townsend, the battalion commander, and Surgeon Cary D. Langhorne, the regimental surgeon of the 2nd Seaman Regiment. During the fighting, Vermont's force suffered one fatality, a private from her Marine detachment, killed on the 23rd. But for a visit to Tampico, Mexico, from 21 September to 10 October, Vermont remained in that Mexican port into late October.

Over the next two and one-half years, Vermont maintained her schedule of operations off the eastern seaboard of the United States, ranging from Newport to Guantanamo Bay, before she lay in reserve at Philadelphia from 1 October to 21 November 1916. Vermont subsequently supported the Marine Corps Expeditionary Force in Haiti from 29 November 1916 to 5 February 1917 and then conducted battle practices out of Guantanamo Bay. She ultimately returned to Norfolk on 29 March 1917.

On 4 April 1917, Vermont entered the Philadelphia Navy Yard for repairs. Two days later, the United States declared war on Germany. The battleship emerged from the yard on 26 August 1917 and sailed for Hampton Roads for duty as an engineering training ship in the Chesapeake Bay region. She performed that vital function for almost the entire duration of hostilities, completing the assignment on 4 November 1918, a week before the armistice stilled the guns of World War I.

Her service as a training ship during the conflict had been broken once in the spring of 1918 when she received the body of the late Chilean ambassador to the United States on 28 May 1918; embarked the American Ambassador to Chile, the Honorable J. H. Shea, on 3 June; and got underway from Norfolk later that day.

The battleship transited the Panama Canal on the 10th, touched at Port Tongoy, Chile, on the 24th; and arrived at Valparaíso on the morning of 27 June.

There, the late ambassador's remains were accompanied ashore by Admiral William B. Caperton and Ambassador Shea. Departing that port on 2 July, Vermont visited Callao, Peru, on the 7th, before retransiting the Panama Canal and returning to her base in the York River.

Vermont entered the Philadelphia Navy Yard on 5 November and was there converted to a troop transport. She subsequently sailed from Norfolk on 9 January 1919 on the first of four round-trip voyages, returning "Doughboys" from "over there." During her time as a transport, the battleship carried some 5,000 troops back to the United States, completing her last voyage on 20 June 1919.

Prepared at the Philadelphia Navy Yard for inactivation, Vermont departed the east coast on 18 July, sailing from Hampton Roads on that day, bound for the west coast. After transiting the Panama Canal, the battleship visited San Diego, San Pedro, Monterey, and Long Beach, California; Astoria, Oregon; and San Francisco, California, before reaching the Mare Island Navy Yard, Vallejo, California, on 18 September. There, the battleship was decommissioned on 30 June 1920. She was subsequently reclassified as BB-20 on 17 July of that same year.

Vermont remained inactive at Mare Island until her name was struck from the Navy list on 10 November 1923. She was then sold for scrapping on 30 November of the same year in accordance with the Washington Treaty limiting naval armaments.

BB-21 • USS KANSAS

(BB-21: displacement 16,000; length 456 feet 4 inches; beam 76 feet 10 inches; draft 24 feet 6 inches; speed 18 knots; complement 880; armament 4 12-inch guns, 8 8-inch guns, 12 3-pounders, 2 1-pounders, 2 .30 caliber machine guns, 4 21-inch torpedo tubes; class Vermont)

The second Kansas (BB-21) was launched by New York Shipbuilding Corp., Camden, N.J., on 12 August 1905; sponsored by Miss Anna Hoch, daughter of the Governor of Kansas; and commissioned at Philadelphia Navy Yard on 18 April 1907, with Captain Charles B. Vreeland in command.

The new battleship departed Philadelphia on 17 August 1907 for shakedown training out of Provincetown, Mass., and returned home for alterations on 24 September. She joined the "Great White Fleet" at Hampton Roads on 9 December and passed in review before President Theodore Roosevelt while getting underway on the first leg of the fleet's historic world cruise. The American ships arrived at Port-of-Spain, Trinidad, on 23 December

and six days later got underway for Rio de Janeiro. From there, they sailed south along the east coast of South America and transited the perilous Straits of Magellan in open order. Turning north, the fleet visited Valparaiso, Chile, and Callao Bay, Peru, en route to Magdalena Bay, Mexico, for a month of target practice.

The "Great White Fleet" reached San Diego on 14 April 1908, and moved on to San Francisco on 7 May. Exactly two months later, the spotless warships sortied through the Golden Gate and headed for Honolulu. From Hawaii, they set course for Auckland, New Zealand, greeted as heroes upon arrival on 9 August. The fleet made Sydney on 20 August and, after enjoying a week of the most warm and cordial hospitality, sailed to Melbourne, where they were welcomed with equal graciousness and enthusiasm.

Kansas had her last glimpse of Australia on 19 September upon leaving Albany for ports in the Philippine Islands, Japan, and Ceylon before transiting the Suez Canal. She departed Port Said, Egypt, on 4 January 1909, for a visit to Villefranche, France, and then staged with the combined "Great White Fleet" at Gibraltar and departed for home on 6 February. She again passed in review before President Roosevelt as she entered Hampton Roads on 22 February, ending a widely acclaimed voyage of goodwill subtly but effectively demonstrating American strength to the world.

A week later, Kansas entered the Philadelphia Navy Yard for overhaul. Repairs completed on 17 June, the battleship began a period of maneuvers, tactical training, and battle practice which lasted almost until the close of the following year. With the 2nd Battleship Division, she sailed on 15 November 1910 for Europe, visiting Cherbourg, France, and Portland, England, before returning to Hampton Roads via Cuba and Santo Domingo. She again departed Hampton Roads on 8 May 1911 for Scandinavia, visiting Copenhagen, Stockholm, Cronstadt, and Kiel before returning to Provincetown, Mass., on 13 July. She engaged in fleet tactics south to the Virginia capes before entering the Norfolk Navy Yard on 3 November for overhaul.

Early in 1912, she began several months of maneuvers out of Guantanamo Bay and then returned to Hampton Roads to serve as one of the welcoming units for the German Squadron which visited there from 28 May to 8 June and New York from 8 to 13 June.

The battleship embarked Naval Academy Midshipmen at Annapolis on 21 June for a summer

practice cruise which took her, among other ports of call along the Atlantic seaboard, to Baltimore during the Democratic National Convention which nominated Woodrow Wilson. After debarking her midshipmen at Annapolis on 30 August, she sailed from Norfolk on 15 November for a training cruise in the Gulf of Mexico. She returned to Philadelphia on 21 December to enter the Navy Yard for overhaul.

Back in top shape on 5 May 1913, Kansas operated on the East Coast until she stood out of Hampton Roads on 25 October, bound for Genoa, Italy. From there, she proceeded to Guantanamo Bay en route to the coast of Mexico to operate off Vera Cruz and Tampico, watching out for American interests in that land then troubled by revolutionary unrest as rival factions struggled to attain and hold power. She returned to Norfolk on 14 March 1914, and entered the Philadelphia Navy Yard for overhaul on 11 April.

Kansas departed Norfolk on 1 July with the body of the Venezuelan Minister to the United States, arriving at La Guaira on 14 July. Then she returned to the Mexican coast to patrol off Tampico and Vera Cruz, supporting the A.E.F. which had landed there. She departed Vera Cruz on 29 October to investigate reports of unstable conditions at Port-au-Prince, Haiti, where she arrived on 3 November. The battleship stood out of Port-au-Prince on 1 December and reached Philadelphia a week later. Maneuvers off the East Coast and out of Guantanamo Bay occupied her until she entered the Philadelphia Navy Yard for overhaul on 30 September 1916.

Kansas was still in that yard on 6 April 1917 when the United States entered World War I. She arrived in York

River from Philadelphia on 10 July and became a unit of the 4th Battleship Division, spending the remainder of the war as an engineering training ship in Chesapeake Bay, occasionally making escort and training cruises to New York. After the Armistice, she made five voyages to Brest, France, to embark and return veterans home.

She was overhauled at the Philadelphia Navy Yard from 29 June 1919 to 17 May 1920. Three days later, she arrived at Annapolis where she embarked midshipmen and sailed on 5 June for a practice cruise to Pacific waters, transiting the Panama Canal to visit Honolulu, Seattle, San Francisco, and San Pedro. She departed the latter port on 11 August, transited the canal, and visited Guantanamo Bay before returning to Annapolis on 2 September.

Proceeding to Philadelphia, Kansas became the flagship of Rear Admiral Charles F. Hughes, Commander of Battleship Division 4, Squadron 2, and future Chief of Naval Operations. She sailed for Bermuda on 27 September and was inspected by the Prince of Wales at Grassy Bay, Bermuda, on 2 October. Two days later, she was underway for the Panama Canal and Samoa. She was in Pago Pago, Samoa, on 11 November when Captain Waldo Evans became Governor of American Samoa. After visiting Hawaiian ports and transiting the Panama Canal, she cruised in the Caribbean and the Panama Canal before returning to Philadelphia on 7 March 1921.

Kansas embarked midshipmen at Annapolis and sailed on 4 June 1921 with three other battleships, bound for Christiana, Norway; Lisbon, Portugal; Gibraltar; and Guantanamo Bay. She returned on 28 August to debark her midshipmen before visiting New York from 3 to 19 September. She entered the Philadelphia Navy Yard on 20 September and was decommissioned on 16 December. Her name was struck from the Navy List on 24 August 1923, and she was sold for scrap in accordance with the Washington Treaty limiting naval armament.

BB-22 • USS MINNESOTA

(BB-22: displacement 10,000; length 456'4"; beam 76'10"; draft 24'6" (mean); speed 18 knots; complement 880; armament 4 12-inch guns, 8 8-inch guns, 12 7-inch guns, 20 3-inch guns, 12 3-pounders, 4 21-inch torpedo tubes; class Connecticut)

The second USS Minnesota (BB-22) was laid down by the Newport News Shipbuilding Co., Newport News, Virginia, on October 27, 1903; launched on April 8, 1905, sponsored by Miss Rose Marie Schaller; and commissioned on March 9, 1907, with Captain J. Hubbard in command.

Following her shakedown off the New England coast, Minnesota was assigned to duty in connection with the Jamestown Exposition in Jamestown, Virginia, from April 22 to September 3, 1907. On December 16, she departed Hampton Roads as one of the 16 battleships sent by President Theodore Roosevelt on a voyage around the world. The cruise of the 'Great White Fleet,' lasting until February 22, 1909, served as a deterrent

to possible hostilities in the Pacific, raised American prestige as a global naval power, and, most importantly, impressed upon Congress the need for a strong navy and a thriving merchant fleet to keep pace with the United States' expanding international interests and far-flung possessions.

Returning from her world cruise, Minnesota resumed operations with the Atlantic Fleet. Over the next three years, she primarily operated along the East Coast, with one brief deployment to the English Channel. In 1912, her schedule began to involve her more in inter-American affairs. During the first half of that year, she cruised in Cuban waters and was stationed at Guantanamo Bay from June 7 to 22, supporting actions aimed at establishing order during the Cuban insurrection. The following spring and summer, she cruised in Mexican waters. In 1914, she twice returned to Mexican waters (January 26 to August 7 and October 11 to December 19) as the country continued in political turmoil. In 1915, she resumed East Coast operations, with occasional cruises to the Caribbean area, which she continued until November 1916 when she became the flagship of the Reserve Force, Atlantic Fleet.

On April 6, 1917, as the United States entered World War I, Minnesota rejoined the active fleet at Tangier Sound, Chesapeake Bay, and was assigned to Division 4, Battleship Force. During World War I, she served as a gunnery and engineering training ship, cruising off the middle Atlantic seaboard until September 27, 1918. On that date, 20 miles from Fenwick Island Shoal Lightship (38°11' N.; 74°41' W.), she struck a mine, apparently laid by the German submarine U-117. Suffering serious damage to the starboard side but with no loss of life, she managed to reach Philadelphia, where she underwent five months of repairs. On March 11, 1919, she returned to sea as a unit of the Cruiser and Transport Force. Assigned to that force until July 23, she completed three round trips to Brest, France, returning over 3,000 veterans to the United States.

Primarily employed thereafter as a training ship, Minnesota took midshipmen on two summer cruises (1920 and 1921) before being decommissioned on December 1, 1921. Struck from the Naval Register the same day, she was dismantled at the Philadelphia Navy Yard, and on January 23, 1924, was sold for scrap.

BB-23 • USS MISSISSIPPI

(BB-23: dp. 13,000 n.; l. 382'; b. 77'; dr. 24'8"; s. 17 k.; cgl. 744; a. 4 12", 8 8", 8 7", 12 3", 2 21" tt.; cl. Mississippi)

The second ship named USS Mississippi (BB-23) was laid down on 12 May 1904 by William Cramp & Sons in Philadelphia, Pennsylvania. She was launched on 30 September 1905, sponsored by Miss M. C. Money, the daughter of Senator H. P. Money of Mississippi, and commissioned at the Philadelphia Navy Yard on 1 February 1908, with Capt. J. C. Fremont in command.

Following her shakedown off the coast of Cuba from 13 February to 13 March 1908, the new battleship returned to Philadelphia for final outfitting. On 1 July, she operated along the New England coast until returning to Philadelphia on 10 September. The warship next set sail on 16 January 1909 to represent the United States at the inauguration of the President of Cuba in Havana, from 25 to 28 January. Mississippi remained in the Caribbean until 10 February, sailing that day to join the "Great White Fleet" as it returned from its world cruise. Reviewed by President Theodore Roosevelt on Washington's Birthday, she returned to the Caribbean on 1 March.

The ship departed Cuban waters on 1 May for a cruise up the Mississippi River, the river for which she was named. Calling at major ports along this great inland waterway, she arrived in Natchez on 20 May, and then proceeded five days later to Horn Island, where she received a silver service from the State of Mississippi. Returning to Philadelphia on 7 June, the battleship operated off the New England coast until sailing on 5 January 1910 for winter exercises and war games out of Guantanamo Bay. Departing on 24 March for Norfolk, she operated off the east coast until fall, visiting numerous large ports, serving as a training ship for the Naval Militia, and engaging in maneuvers and exercises.

On 1 November, she departed Philadelphia for a fleet rendezvous at Gravesend Bay, England, on 16

November, and then sailed for Brest, France, arriving on the 9th of December. On 30 December, Mississippi set course for Guantanamo Bay for winter maneuvers until 13 March 1911.

Returning to the United States, the battleship operated off the Atlantic coast, alternating bases between Philadelphia and Norfolk for the next year and two months. She served as a training ship and conducted operational exercises. Departing Tompkinsville, New York, on 26 May 1912 with a detachment from the 2nd Marine Regiment on board to protect American interests in Cuba, she landed her Marine detachment at El Cuero on 19 June, remaining in Guantanamo Bay until 5 July, when she sailed for home.

After exercises with the 4th Battleship Division off New England, she returned to the Philadelphia Navy Yard and was placed in the 1st Reserve on 1 August 1912.

Mississippi remained in the Atlantic Reserve Fleet at Philadelphia until 30 December 1913 when she was detached for duty as an aeronautic station ship at Pensacola, Florida. Departing on 6 January 1914, she arrived on 21 January, transporting equipment for the establishment of a naval air station. In Pensacola, she supported the early naval aviators in rebuilding the old naval base, laying the foundation for what would become the largest and most famous American naval air station.

With the outbreak of fighting in Mexico, Mississippi sailed on 21 April to Vera Cruz, arriving on the 24th with the first detachment of naval aviators to enter combat. Serving as a floating base for the fledgling seaplanes and their pilots, the warship launched nine reconnaissance flights over the area during an 18-day period, with the last flight on 12 May. A month later, the battleship departed Vera Cruz for Pensacola, serving as a station ship there from 15 to 28 June before sailing north to Hampton Roads, where she transferred her aviation gear to the armored cruiser North Carolina (CA-12) on 3 July.

On 10 July, Mississippi shifted to Newport News to prepare for transfer to the Greek Government. She was decommissioned at Newport News on 21 July 1914 and turned over to the Royal Hellenic Navy the same day. Renamed Lemnos, the battleship served for the next 17 years as a coast defense vessel. She was sunk in an air attack by German bombers in Salamis harbor in April 1941, and after World War II, her hull was salvaged for scrap.

BB-24 • USS IDAHO

(BB-24: dp. 13,000 n.; l. 382'; b. 77'; dr. 24'8"; s. 17 k.; cgl. 744; a. 4 12", 8 8", 8 7", 12 3", 2 21" tt.; cl. Mississippi)

The second ship named Idaho (BB-24) was launched by William Cramp & Sons, Philadelphia, on 9 December 1905, sponsored by Miss Louise Gooding, daughter of the Governor of Idaho, and commissioned at the Philadelphia Navy Yard on 1 April 1908 with Captain S.W.B. Diehl in command.

The new battleship conducted a shakedown cruise to Cuba in April-May 1908, and after a visit to Panama, returned to Philadelphia for alterations. The ship participated in a large naval review in Hampton Roads on 22 February 1909, celebrating the return of the Great White Fleet from its around-the-world cruise. In March, she returned to the Caribbean for maneuvers, continuing to take part in training operations until October 1910. Idaho sailed on 29 October for exercises in British and French waters and upon her return

participated in gunnery exercises in Chesapeake Bay from 19 to 23 March 1911.

Idaho sailed from Philadelphia on 4 May 1911 for a cruise up the Mississippi River to Louisiana ports. She then steamed to the east coast of Florida for battleship maneuvers and continued to operate off the coast and in the Caribbean until entering the reserve at Philadelphia on 27 October 1913. There she remained until 9 May 1914, when the ship sailed to the Mediterranean with midshipmen for at-sea training. After visiting various ports in North Africa and Italy and carrying out a rigorous training program, Idaho arrived in Villefranche on 17 July 1914, transferred her crew to Maine, and decommissioned on 30 July. She was then turned over to the government of Greece, where she served as the coastal defense ship Kilkis until being sunk in Salamis harbor by German aircraft in April 1941.

BB-25 • USS NEW HAMPSHIRE

(BB-25: displacement. 16 000, length. 456'4", beam. 76'10", draft. 24'6", speed 18 k., complement. 850; armor 4 12", 8 8", 12 7", 20 3", 2 1-pdrs., 4 21" tt.;cl. Connecticut)

The second New Hampshire (BB-25) was laid down on 1 May 1905 by New York Shipbuilding Corp., Camden, N.J. Launched 30 June 1906, sponsored by Miss Hazel E. MeLane, daughter of Governor John MeLane of New Hampshire, and commissioned 19 March 1908, Capt. Cameron M. Winslow in command.

After fitting out at New York, New Hampshire carried a Marine Expeditionary Regiment to Colon, Panama, 20 26 June 1908, then made ceremonial visits to Quebec, Portsmouth, New York, and Bridgeport. Overhaul at New York and Caribbean exercises were followed by participation in the Naval Review by President Theodore Roosevelt in Hampton Roads on 22 February 1909, welcoming home the "Great White Fleet." Through the next year and a half she exercised along the east coast and in the Caribbean, then departed Hampton Roads 1 November 1910 with the Second Battleship Division for Cherbourg, France and Weymouth, England. Leaving England 30 December, she returned to the Caribbean until arriving in Norfolk 10 March 1911 to prepare for a second European cruise which took her to Scandinavian, Russian and German ports. The squadron returned to New England waters 13 July

New Hampshire trained Naval Academy midshipmen off New England in the next two summers, and patrolled off strife-torn Hispaniola in December 1912. From 14 June 1913 until 29 December, she similarly protected American interests along the Mexican coast, to which she returned 15 April 1914 to support the occupation of Vera Cruz. New Hampshire sailed north 21 June, was overhauled at Norfolk, and exercised along the east coast and in the Caribbean until returning to Vera Cruz in August 1915.

Arriving in Norfolk on 30 September 1915, New Hampshire operated in northern waters until 2 December 1916, when she sailed for Santo Domingo, where her commanding officer took part in the government of the revolt-torn country. She returned to Norfolk in February 1917 for overhaul, where she lay when the United States entered World War I. For the next year and a half, she trained gunners and engineers in northern coastal waters and, on 15 September, began the first of two convoy escort missions, guarding transports from New York to a rendezvous point off the French coast. On 24 December, she sailed on the first of four voyages, bringing veterans home from France to east coast ports. This duty was completed on 22 June 1919; she was overhauled at Philadelphia, then on 5 June 1920, sailed with Academy midshipmen and embarked for a cruise through the Panama Canal to Hawaii and west coast ports. She returned to Philadelphia 11 September.

New Hampshire served as flagship for the special naval force in Haitian waters from 18 October to 12 January 1921 and on 25 January sailed with the remains of Swedish Minister Wilhelm Ekengren for Stockholm arriving 14 February. She called also at Kiel and Gravesen] before returning to Philadelphia 24 March. There she decommissioned 21 May 1921.

She was sold for scrapping 1 November 1923 in accordance with the Washington Treaty for the Limitation of Naval Armaments.

Construction of New Hampshire (BB-70), a battleship to be built by New York Navy Yard, Brooklyn, N.Y., was cancelled 21 July 1943.

BB-26 • USS SOUTH CAROLINA

(BB-26: displacement 16,000 (normal); length 452'9"; beam 80'2½" (waterline); draft 24'6½" (mean); speed 18.86 knots (trial); complement 751; armament 8 12-inch guns, 22 3-inch guns, 2 3-pounder guns, 2 21-inch torpedo tubes; class South Carolina).

The fourth South Carolina (Battleship No. 26) was laid down on 18 December 1906 at Philadelphia by William Cramp & Sons. It was launched on 1 July 1908, sponsored by Miss Frederica Ansel, and commissioned on 1 March 1910 with Captain Augustus F. Fechteler in command.

South Carolina departed Philadelphia on 6 March for a shakedown cruise, visiting the Danish West Indies and Cuba, then Charleston, S.C., from 10 to 15 April. After conducting trials off the Virginia Capes and Provincetown, Mass., the battleship visited New York City on 17 and 18 June for a reception for former President Theodore Roosevelt. Voyage repairs at Norfolk, naval militia training duty, and Atlantic Fleet maneuvers off

Provincetown and the Virginia Capes occupied her until November. Between 1 November 1910 and 12 January 1911, she voyaged to Europe and back with the 2nd Battleship Division, visiting Cherbourg, France, and Portland, England. Upon returning to Norfolk, she underwent repairs, then conducted tactical training and maneuvers off the New England coast.

Following a short visit to New York, she sailed east with the 2nd Battleship Division to Copenhagen, Denmark, Stockholm, Sweden, and Kronstadt, Russia. During the return from Kronstadt, she reached Kiel, Germany, on 21 June in time for the Kiel Yachting Week, hosted by Kaiser Wilhelm II. On 13 July 1911, she arrived off Provincetown, Mass., for battle practice along the coast to the Chesapeake Bay.

Late in 1911, she participated in a naval review in New York and maneuvers with the 1st Squadron out of Newport, R.I. On 3 January 1912, she left New York for winter operations near Guantanamo Bay, Cuba, returning to Norfolk on 13 March. Until late June, she cruised the East Coast as far north as Newport. In June, she participated in welcome receptions at Hampton Roads and New York for the visiting German Squadron, comprised of the battle cruiser Moltke and two cruisers, Bremen and Stettin. On 30 June, she entered the Norfolk yard for an overhaul.

Just over three months later, she sailed to New York for a four-day visit, from 11 to 15 October. This was followed by a month of exercises off the coasts of New England and the Virginia Capes. From mid-November to mid-December, South Carolina steamed with the Special Service Division, visiting Pensacola, New Orleans, Galveston, and Vera Cruz, Mexico. She returned to Norfolk on 20 December, staying there until 6 January 1913, when she sailed to Colón, Panama, to observe the newly-completed Panama Canal. After maneuvers near Guantanamo Bay, she returned to Norfolk on 22 March and cruised north to Newport, stopping in New York from 28 to 31 May for the dedication of a memorial to the battleship Maine.

After training midshipmen in the Virginia Capes area, South Carolina embarked on a 16-month mission in the Gulf of Mexico and the Caribbean Sea. From late June to mid-September 1913, she patrolled the eastern coast of Mexico, protecting American interests in Tampico

and Vera Cruz. After an overhaul in Norfolk from late September 1913 to early January 1914, she headed to Culebra Island for maneuvers.

On 28 January, the battleship landed marines at Port-au-Prince, Haiti, to guard the United States legation and establish a field radio station amid political turmoil. She left Port-au-Prince on 14 April after some stability was restored under General Orestes Zamor, the new Haitian President. After coaling at Key West, she sailed to Vera Cruz, sending a landing force ashore to join the city's occupation for a month. South Carolina spent the troubled summer of 1914 monitoring conditions in Santo Domingo and Haiti.

By her return to Norfolk on 24 September, World War I had been underway for two months. On 14 October, she entered the Philadelphia yard for revitalization, reemerging on 20 February 1915. She headed south for battle practice near Cuba, significant due to the diplomatic crisis following Germany's declaration of the waters around England as a war zone. Despite the sinking of Lusitania, the U.S. remained neutral, and South Carolina continued routine operations of exercises, summer operations, and periodic repairs until the U.S. joined the war in April 1917.

During the war, South Carolina primarily operated along the East Coast. On 9 September 1918, she escorted

a convoy to France, returning to the U.S. after a brief period. After the Armistice on 11 November 1918, she resumed gunnery training service.

From mid-February to late July 1919, South Carolina made four round trips between the U.S. and Brest, France, transporting over 4,000 World War I veterans. After an overhaul at Norfolk Navy Yard, she embarked midshipmen at Annapolis for a Pacific cruise, departing on 5 June 1920. She transited the Panama Canal, visited Hawaii and the West Coast, and returned to Annapolis on 2 September via San Diego and the canal. She then sailed to Philadelphia, where she remained for seven months.

In April 1921, South Carolina cruised to Culebra Island in the West Indies for training, then operated in Chesapeake Bay. On 29 May, she embarked midshipmen at Annapolis for a summer training cruise, calling at Christiania, Norway, and Lisbon, Portugal, before heading to Guantanamo Bay. She debarked the midshipmen at Annapolis on 30 August and steamed to Philadelphia, arriving the next day. South Carolina was decommissioned at Philadelphia on 15 December 1921 and remained there until her name was struck from the Navy list on 10 November 1923. Her hulk was sold for scrap on 24 April 1924, in accordance with the Five-Power Naval Treaty of Washington.

BB-27 • USS MICHIGAN

(BB-27: displacement 16,000; length 452'9"; beam 80'3"; draft 24'6"; speed 18.5 knots; complement 869; armament 8 12-inch guns, 22 3-inch guns, 4 1-pounders, 2 .30 caliber machine guns, 2 21-inch torpedo tubes; class South Carolina).

The second USS Michigan (BB-27) was laid down on December 17, 1908, by the New York Shipbuilding Co., Camden, New Jersey, launched on May 26, 1908, sponsored by Mrs. F.W. Brooks, daughter of Secretary of the Navy Truman Newberry, and commissioned on January 4, 1910, with Captain N.R. Usher in command.

Assigned to the Atlantic Fleet, Michigan conducted shakedown off the East Coast and in the Eastern Caribbean until July 7, 1910. Departing New York Harbor on July 29, the battleship then steamed along the New England and Middle Atlantic Coasts for maneuvers. On November 2, she departed Boston, Massachusetts, for a training cruise to Western Europe. After visiting Portland,

England, she arrived in Cherbourg, France, on December 4. She sailed on December 30 for the Caribbean, touched Guantanamo Bay, Cuba, on January 10, 1911, and reached Norfolk on the 14th.

Michigan operated along the Atlantic coast until setting out from the Virginia Capes on November 15, 1912, for a cruise to the Gulf of Mexico. After visiting Pensacola, New Orleans, and Galveston, she arrived in Veracruz on December 12. She headed for home two days later and reached Hampton Roads on the 20th. She operated along the East Coast until departing Quincy, Massachusetts, on July 8 for the Gulf Coast of Mexico to protect American interests endangered by civil strife in Mexico. The battleship anchored off Tampico on the 11th and remained alert off the Mexican coast until sailing for New York on January 13, 1914, reaching Brooklyn Navy Yard on the 20th.

She began a run from Norfolk to Guantanamo Bay, Cuba, on February 14 and returned to Hampton

Roads in March. Underway again on April 16, she joined American forces upholding American honor at Veracruz. Reaching that troubled Mexican city on April 22, she landed a battalion of Marines as part of the main occupation force, then operated off the Mexican coast, heading home on June 20 and entering the Delaware Capes on the 26th.

Michigan next put to sea on October 21, 1914, and from that time until the eve of America's entry into World War I operated out of various ports on the Eastern seaboard. Assigned to Battleship Force 2 on April 6, 1917, the warship escorted convoys, trained recruits, and engaged in fleet maneuvers and battle practice. On January 15, 1918, while steaming in formation with the fleet off Cape Hatteras, Michigan's foremast buckled and was carried away over the port side as the battleship lurched violently in the trough of a heavy sea. Six men were killed and 13 injured, five seriously, in this accident. Michigan proceeded to Norfolk where the next day she transferred her casualties to Solace (AH-5). On the 22nd, she entered Philadelphia Navy Yard for repairs. In early April, she resumed operations off the East Coast and trained gunners in Chesapeake Bay until World War I ended.

Ordered to duty with the Cruiser and Transport Force in late December 1918, the battleship made two voyages to Europe, from January 28 to March 3 and from March 18 to April 26, 1919, returning 1,052 troops to the United States.

Following an overhaul at Philadelphia during May and June, Michigan resumed training exercises in the Atlantic until August 6, when she was placed in limited commission at Philadelphia Navy Yard. She next put to sea on May 19, 1920, sailing to Annapolis to embark midshipmen for a training cruise through the Panama Canal to Honolulu, Hawaii, arriving on June 3. The cruise continued to major West Coast naval bases and Guantanamo Bay before the battleship returned home on September 2. She returned to Philadelphia on September 5, and was placed in ordinary until sailing on April 4, 1921, for the Caribbean. Returning to Hampton Roads on April 23, she reached Annapolis on May 28 to begin her second midshipmen training cruise. She got underway on June 4 for Europe, visiting Kristiania, Norway, Lisbon, Portugal, and Gibraltar, and returning via Guantanamo to Hampton Roads on August 22. The veteran battleship put to sea on August 31 to make her final cruise up the Delaware River to Philadelphia, arriving on September 1. Michigan decommissioned at Philadelphia Navy Yard on February 11, 1922, and was stricken from the Navy list on November 10, 1923. In accordance with the treaty limiting naval armaments, she and four other battleships were scrapped by the Philadelphia Navy Yard during 1924. Materials from their hulls were sold to four different foundries.

BB-28 • USS DELAWARE

(BB-29: t. 20,000; l. 518'9"; b. 85'3"; dr. 26'11", s. 21 k.; cpl. 933; a. 10 12", 14 5" 4 3-pdrs., 2 21" tt.; cl. Delaware)

The sixth shipped named Delaware (BB-28) was launched on 6 February 1909 by the Newport News Shipbuilding Co., Newport News, Virginia, sponsored by Mrs. A.P. Cahall, niece of the Governor of Delaware, and commissioned on 4 April 1910 with Captain C.A. Gove in command.

After visiting Wilmington, Delaware, from 3 to 9 October 1910 to receive a gift of a silver service from the state, Delaware sailed from Hampton Roads on 1 November with the First Division, Atlantic Fleet. The ship visited Weymouth, England, and Cherbourg, France, and after battle practice at Guantanamo Bay, Cuba, returned to Norfolk on 18 January 1911. She departed on 31 January to carry the remains of Chilean Minister Cruz to Valparaíso, sailing via Rio de Janeiro, Brazil, and Punta Arenas, Chile. Returning to New York

on 6 May, she sailed on 4 June for Portsmouth, England, where from 19 to 28 June she took part in the fleet review accompanying the coronation of King George V.

From 1912 to 1917, in operations with the Fleet, Delaware participated in exercises, drills, and torpedo practice at Rockport and Provincetown, Massachusetts. She engaged in special experimental firing and target practice at Lynnhaven Roads, trained in Cuban waters, and provided summer training for midshipmen. She passed before President Taft and the Secretary of the Navy in the Naval Review of 14 October 1912 and visited Villefranche, France, in 1913 with battleships Wyoming (BB-32) and Utah (BB-31). In 1914 and 1915, she cruised off Veracruz to protect American lives and property during the political disturbances in Mexico.

With the outbreak of World War I in Europe, Delaware returned from winter maneuvers in the Caribbean to train armed guard crews and engineers and to prepare the Fleet for war. On 25 November

1917, she sailed from Lynnhaven Roads with Division 9 for Scapa Flow, Scotland. After encountering bad weather in the North Atlantic, she joined the 6th Battle Squadron, British Grand Fleet, on 14 December for exercises.

On 6 February 1918, the 6th Battle Squadron escorted a group of merchant ships to Norway. Near Stavanger, Delaware evaded two submarine attacks. The squadron returned to Scapa Flow on 10 February. Delaware participated in two more convoy voyages in March and April, then sailed with the Grand Fleet on 24 April to reinforce the 2nd Battle Cruiser Squadron. The vessels of the advance screen made contact with the enemy, but no action ensued.

From 30 June to 2 July 1918, the 6th Battle Squadron screened American ships laying the North Sea mine barrage. On 22 July, George V inspected the ships of the Grand Fleet at Rosyth, Scotland. Delaware was relieved by Arkansas (BB-33) on 30 July and sailed for Hampton Roads, arriving on 12 August.

Delaware remained at York River until 12 November 1918, then sailed to Boston Navy Yard for an overhaul. On 11 March 1919, she rejoined the Fleet in Cuban waters for exercises. Returning to New York on 14 April, she continued to operate in division, squadron, and fleet maneuvers and participated in the Presidential Fleet Review at Hampton Roads on 28 April 1921. She made two midshipmen practice cruises, one to Colón, Martinique, and other ports in the Caribbean and to Halifax, Nova Scotia, between 6 June and 31 August 1922, and a second to Europe, visiting Copenhagen, Greenock, Cadiz, and Gibraltar between 9 July and 29 August 1923.

Delaware entered Norfolk Navy Yard on 30 August 1923, and her crew was transferred to Colorado (BB-45), a newly commissioned battleship. Moving to Boston Navy Yard in September, she was stripped of warlike equipment and decommissioned on 10 November 1923. Delaware was sold on 5 February 1924 and scrapped in accordance with the Washington Treaty on the limitation of armaments.

BB-29 • USS NORTH DAKOTA

(BB-29: t. 20,000; l. 518'9"; b. 85'3"; dr. 26'11", s. 21 k.; cpl. 933; a. 10 12", 14 5" 4 3-pdrs., 2 21" tt.; cl. Delaware)

North Dakota (BB-29) was laid down on 16 December 1907 by Fore River Shipbuilding Co., Quincy, Mass.; launched on 10 November 1908; sponsored by Miss Mary Benton, and commissioned at Boston on 11 April 1910, Cmdr. Charles P. Plunkett in command.

In her first years, North Dakota operated with the Atlantic Fleet in maneuvers along the East Coast and in the Caribbean. She sailed on 2 November 1910 for her first Atlantic crossing, visiting England and France prior to winter-spring maneuvers in the Caribbean. In the summers of 1912 and 1913, she carried Naval Academy midshipmen for training in New England waters, and on 1 January 1913, she joined the honor escort for Natal as the Brazilian ship entered New York harbor with the body of the late Whitelaw Reid, United States Ambassador to Brazil.

As Mexican political disturbances strained relations with the United States, North Dakota sailed for Vera Cruz, where she arrived on 26 April 1914, five days after American sailors had occupied the city. She cruised the coast of Mexico to protect Americans and their interests until a more stable government took office, and returned to Norfolk on 16 October. An even more intensive program of training was taken up by the Atlantic Fleet as war threatened, and North Dakota was in Chesapeake Bay for gunnery drills when the United States entered World War I.

Throughout the war, North Dakota operated in the York River, Va., and out of New York training gunners and engineers for the expanding fleet. Then, on 13 November 1919, she stood out of Norfolk to carry home the remains of the late Italian Ambassador to the United States. While in the Mediterranean, she called at Athens, Constantinople, Valencia, and Gibraltar before returning to the Caribbean for the annual spring maneuvers. In the summer of 1921, she took part in the Army-Navy bombing tests off the Virginia Capes in which the German warships Frankfurt and Ostfriesland were sunk to demonstrate the potentialities of airpower. She interrupted fleet operations during the next two summers to again cruise with midshipmen, contributing to the future strength of the Navy by educating its officers-to-be. The cruise of 1923 took her to Scandinavia, Scotland, and Spain.

North Dakota decommissioned at Norfolk on 22 November 1923. Her name was struck from the Navy List on 7 January 1931, and she was sold for scrapping on 16 March 1931.

BB-30 • USS FLORIDA

(Battleship No. 30: dp. 21,826 (n.); l. 621'6"; b. 88'3";dr. 28'4" (mean); s. 20.76 k., cpl. 1,041; a. 10 12",16 6", 2 21" tt.; cl. Florida)

The fifth Florida (BB-30) was launched on 12 May 1910 by the New York Navy Yard, sponsored by Miss E.D. Fleming, daughter of a former Florida governor, and commissioned on 15 September 1911 with Captain H.S. Knapp in command.

After extensive training in the Caribbean and Maine coastal waters, Florida arrived in Hampton Roads, VA, on 29 March 1912 to join the Atlantic Fleet as flagship of Division 1. Regularly scheduled exercises, maneuvers, fleet training, target practice, and midshipmen training cruises took the new battleship to many East Coast ports and into Caribbean waters. Early in 1914, tension heightened between the United States and factions in Mexico, and Florida arrived off Veracruz on 16 February, remaining there during the ensuing occupation. She

steamed to New York in July to resume regular fleet operations and in October was transferred to Division 2.

Following the United States' entry into World War I, Florida completed exercises in the Chesapeake Bay and proceeded with Battleship Division 9 to join the British Grand Fleet at Scapa Flow, Orkney Islands, on 7 December 1917. She participated in the Grand Fleet's maneuvers and evolutions and performed convoy duty with the 6th Battle Squadron through the remainder of the war. She rendezvoused with the Grand Fleet on 20 November 1918 when it met to escort the German High Seas Fleet into the Firth of Forth.

Florida joined the escort for George Washington, with President Woodrow Wilson embarked, as she proceeded into Brest, France, on 12 and 13 December 1918. She participated in the grand Victory Naval Review in the North River, New York City, in late December and then returned to Norfolk on 4 January 1919 to resume peacetime operations. During May, she cruised to the

Azores and took weather observations for the first aerial crossing of the Atlantic achieved that month by Navy seaplanes.

Florida's operations during the remaining years of her career were highlighted by participation in the tercentenary celebration in August 1920 of the Pilgrims' landing at Provincetown, Mass.; a diplomatic voyage to South American and Caribbean ports with Secretary of State R. Lansing embarked; service as flagship for Commander, Control Force, U.S. Fleet; amphibious operations with Marines in the Caribbean; and midshipman training cruises. She was decommissioned at Philadelphia on 16 February 1931 and scrapped under the terms of the London Naval Treaty of 1930.

BB-31 • USS UTAH

(Battleship No. 31: dp. 21,826 (n.); l. 621'6"; b. 88'3";dr. 28'4" (mean); s. 20.76 k., cpl. 1,041; a. 10 12",16 6", 2 21" tt.; cl. Florida)

Utah (Battleship No. 31) was laid down on 9 March 1909 at Camden, N.J., by the New York Shipbuilding Co.; launched on 23 December 1909; sponsored by Miss Mary Alice Spry, daughter of Governor William Spry of Utah, and commissioned at the Philadelphia Navy Yard on 31 August 1911, Capt. William S. Benson in command.

After her shakedown cruise voyage that took her to Hampton Roads, Va.; Santa Rosa Island and Pensacola, Fla.; Galveston, Tex.; Kingston and Portland Bight, Jamaica; and Guantanamo Bay, Cuba, Utah was assigned to the Atlantic Fleet in March 1912. She operated with the Fleet early that spring, conducting exercises in gunnery and torpedo defense, before she entered the New York Navy Yard on 16 April for an overhaul.

Departing New York on 1 June, Utah briefly visited Hampton Roads and then steamed to Annapolis, Md., where she arrived on the 6th. There, she embarked Naval Academy midshipmen and got underway on the 10th for the Virginia capes and the open Atlantic. She conducted a midshipmen training cruise off the New England seaboard well into the summer before disembarking her contingent of officers-to-be back at Annapolis on 24 and 26 August. Soon thereafter, the battleship headed for the Southern Drill Grounds to conduct gunnery exercises.

For a little over two years, the dreadnought maintained that schedule of operations off the eastern seaboard, ranging from the New England coast to Cuban waters. During that time, she made one cruise to European waters, visiting Villefranche, France, from 8 to 30 November 1913.

Utah began the year 1914 at the New York Navy Yard and sailed south on 6 January. After stopping at Hampton

Roads, she reached Cuban waters later in the month for torpedo and small arms exercises. However, due to tension in Mexico, Utah sailed for Mexican waters in early February and reached Vera Cruz on the 16th. She operated off that port until getting underway for Tampico on 9 April with several hundred refugees embarked.

Soon thereafter, it was learned that a German steamship, SS Ypiranga, was bound for Vera Cruz with a shipment of arms and munitions earmarked for the dictator Victoriano Huerta. Utah received orders to search for the ship and put to sea and reached Vera Cruz on the 16th. When it appeared that the shipment might be landed, the Navy took steps to take the customs house at Vera Cruz and stop the delivery. Accordingly, plans were drawn up for a landing at Vera Cruz, to commence on 21 April 1914.

Utah consequently landed her "battalion" 17 officers and 367 sailors under the command of Lt. Guy W. S. Castle, as well as her Marine detachment, which formed part of the improvised "First Marine Brigade," made up of detachments of marines from the other ships that had arrived to show American determination. In the ensuing fighting, in which the men of Utah's bluejacket battalion distinguished themselves, seven won medals of honor. Those seven included Lt. Castle, the battalion commander; company commanders Ens. Oscar C. Badger and Ens. Paul F. Foster, section leaders Chief Turret Captains Niels Drustrup and Abraham DeSomer; Chief Gunner George Bradley; and Boatswain's Mate Henry N. Nickerson.

Utah remained at Vera Cruz for almost two months before returning north to the New York Navy Yard in late June for an overhaul. Over the next three years, the battleship operated on a regular routine of battle practices and exercises from off the eastern seaboard into the Caribbean, as the United States readied its forces for the possible entry of the United States into the worldwide war that broke out in July 1914.

After the United States finally declared war on 6 April 1917, Utah operated in the waters of the Chesapeake Bay as an engineering and gunnery training ship and continued that duty until 30 August 1918, when she sailed for the British Isles with Vice Admiral Henry T. Mayo, Commander in Chief, United States Atlantic Fleet, embarked.

Fears of possible attacks by German heavy units upon the large convoys crossing the Atlantic with troops and munitions for the Western Front prompted the dispatch to European waters of a powerful force of American dreadnoughts to Irish waters. Utah, as part of that movement, reached Berehaven, Bantry Bay, Ireland, on 10 September. There, she became the flagship of Rear Admiral Thomas S. Rodgers, Commander, Battleship Division 6. Until the signing of the armistice on 11 November 1918, Utah, along with the sister ships Oklahoma (Battleship No. 37) and Nevada (Battleship No. 36), operated from Bantry Bay, covering the Allied convoys approaching the British Isles, ready to deal with any surface threat that the German Navy could hurl at the valuable transports and supply ships.

After the cessation of hostilities, Utah visited Portland, England, and later served as part of the honor escort for the transport George Washington (Id. No. 3018), as that ship bore President Woodrow Wilson into the harbor of Brest, France, on 13 December 1918. The following day, Utah turned homeward and reached New York on Christmas Day 1918.

Utah remained at anchor in the North River, off New York City, until 30 January 1919. During that time, she half-masted her colors at 1440 on 7 January due to the death of former President Theodore Roosevelt and, on the 8th, fired salutes at half-hour intervals throughout the day in memory of the great American statesman.

Utah carried out a regular routine of battle practices and maneuvers, ranging from the New England coast to the Caribbean, into mid-1921. During that time, she was classified as BB-31 on 17 July 1920, during the Navy-wide assignment of hull numbers.

Ultimately departing Boston on 9 July 1921, Utah proceeded via Lisbon, Portugal, and reached Cherbourg, France, soon thereafter. There, Utah became the flagship for the United States naval forces in European waters. She "showed the flag" at the principal Atlantic coast ports of Europe and in the Mediterranean until relieved by Pittsburgh (CA 4) in October 1922.

Returning to the United States on 21 October 1922, Utah then became the flagship of Battleship Division (BatDiv) 5, United States Scouting Fleet, and operated with the Scouting Fleet over the next three and a half years.

Late in 1924, Utah was chosen to carry the United States diplomatic mission to the centennial celebration of the Battle of Ayacucho (9 December 1824), the decisive action in the Peruvian struggle for independence. Designated as flagship for the special squadron assigned to represent the United States at the festivities, Utah departed New York City on 22 November 1924 with General of the Armies John J. Pershing, USA, and former congressman, the Honorable F. C. Hicks, embarked, and arrived at Callao on 9 December.

Utah disembarked General Pershing and the other members of the mission on Christmas 1924, so that the general and his mission could visit other South American cities inland on their goodwill tour. Meanwhile, Utah, in the weeks that followed, called at the Chilean ports of Punta Arenas and Valparaiso before she rounded Cape Horn and met General Pershing at Montevideo, Uruguay. Reembarking the general and his party there, the battleship then visited in succession: Rio de Janeiro, Brazil; La Guaira, Venezuela; and Havana, Cuba, before ending her diplomatic voyage at New York City on 13 March 1925.

Utah spent subsequent summers of 1925 and 1926 with the Midshipman Practice Squadron and, after disembarking her midshipmen at the conclusion of

the 1926 cruise, entered the Boston Navy Yard and was decommissioned on 31 October 1926 for modernization. During that period of alterations and repairs, the ship's "cage" mainmast was replaced by a lighter pole mast; she was fitted to burn oil instead of coal as fuel; and her armament was modified to reflect the increased concern over anti-aircraft defense. Interestingly, Utah and her sister ship Florida (BB-30) never received the more modern "tripod" masts fitted to other classes.

Utah was placed back in commission on 1 December 1926 and, after local operations with the Scouting Fleet, departed Hampton Roads on 21 November 1928, bound for South America. Reaching Montevideo on 18 December, she there embarked President-elect and Mrs. Herbert C. Hoover; the Honorable Henry T. Fletcher, Ambassador to Italy; and members of the press. Utah transported the President-elect's party to Rio de Janeiro, Brazil, between 21 and 23 December, and then continued her homeward voyage with Mr. Hoover embarked. En route, the President-elect inspected the battleship's crew while at sea, before the ship reached Hampton Roads on 6 January 1929.

However, Utah's days as a battleship were numbered. Under the terms of the 1922 Washington Naval Treaty, Utah was selected for conversion to a mobile target, in place of the former battleship North Dakota; and, on 1 July 1931, Utah's classification was changed to AG-16. Her conversion, carried out at the Norfolk Navy Yard, included the installation of a radio-control apparatus. After having been decommissioned for the duration of the conversion, Utah was recommissioned at Norfolk on 1 April 1932, Cmdr. Randall Jacobs in command.

Utah departed Norfolk on 7 April to train her engineers in using the new installations and for trials of her radio gear by which the ship could be controlled at varying rates of speed and changes of course maneuvers that a ship would conduct in battle. Her electric motors, operated by signals from the controlling ship, opened and closed throttle valves, moved her steering gear, and regulated the supply of oil to her boilers. In addition, a Sperry gyro pilot kept the ship on course.

Returning to port on 21 April, Utah passed her radio control trials off the Virginia capes on 6 May. On 1 June, Utah ran three hours under radio control, with all engineering stations manned, over the next two days, she made two successful runs, each of four hours duration, during which no machinery was touched by human hands. Observers, however—two in each fire room and two in each boiler room—kept telephone information and recorded data.

Her trials completed, Utah departed Norfolk on 9 June. After transiting the Panama Canal, she reached San Pedro, Calif., on 30 June, reporting for duty with Training Squadron 1, Base Force, United States Fleet. She conducted her first target duty, for cruisers of the Fleet, on 26 July, and later, on 2 August, conducted rehearsal runs for Nevada (BB-36), Utah being controlled from Hovey (DD-208) and Talbot (DD-114).

Over the next nine years, the erstwhile battleship performed a vital service to the fleet as a mobile target, contributing realism to the training of naval aviators in dive, torpedo, and high-level bombing. Thus, she greatly aided the development of tactics in those areas. On one occasion, she even served as a troop transport, embarking 223 officers and men of the Fleet

Marine Force at Sand Island, Midway, for amphibious operations at Hilo Bay, Hawaii, as part of Fleet Problem XVI in the early summer of 1936. She then transported the marines from Hawaii to San Diego, Calif., disembarking them there on 12 June 1935.

That same month, June 1935, saw the establishment of a fleet machine gun school onboard Utah while she continued her mission as a mobile target. The former dreadnought received her first instructors on board in August 1935, and the first students—drawn from the ships' companies of Raleigh (CL-7), Concord (CL-10), Omaha (CL-4), Memphis (CL-13), Milwaukee (CL-5), and Ranger (CV-4)—reported aboard for training on 20 September. Subsequently, during the 1936 and 1937 gunnery year, Utah was fitted with a new quadruple 1.1-inch machine gun mount for experimental test and development by the machine gun school. Some of the first tests of that type of weapon were conducted on board.

Utah—besides serving as a realistic target for exercises involving carrier-based planes—also towed targets during battle practices conducted by the Fleet's battleships and took part in the yearly "fleet problems." She transited the Panama Canal on 9 January 1939 to participate in Fleet Problem XX—part of the maneuvers observed personally by President Franklin D. Roosevelt from the heavy cruiser Houston (CA-30).

After providing mobile target services for the submarines of Submarine Squadron 6 in the late autumn and early winter of 1939, Utah devoted the eight months that followed to special machine gun practices. The following summer, Utah sailed for the Hawaiian Islands, reaching Pearl Harbor on 1 August 1940, and fired advanced anti-aircraft gunnery practice in the Hawaiian operating area until 14 December 1940, when she sailed for the west coast, returning to Long Beach four days before Christmas.

For the next two months, Utah operated as a mobile bombing target off San Clemente Island, Calif., for planes from Patrol Wing 1, and from the carriers Lexington (CV-2), Saratoga (CV-3), and Enterprise (CV-6). Utah returned to Hawaiian waters on 1 April 1941, embarking gunners

for the Advanced Anti-aircraft Gun School, men drawn from West Virginia (BB-48), Oklahoma (BB-37), Colorado (BB-45), Phoenix (CL-46), Nashville (CL-43), Philadelphia (CL-41), and New Orleans (CA-32).

Over the weeks that followed, she trained her embarked gunnery students in control and loading drills for the 5-inch batteries, firing runs on radio-controlled drone targets as well as .50-caliber and 1.1-inch firing on drones and balloons. Utah put into Los Angeles harbor on 20 May and there embarked Fleet Marine Force passengers for transportation to Bremerton, Wash. Putting the marines ashore a week later, the ship entered the Puget Sound Navy Yard on 31 May 1941.

During the ensuing overhaul, Utah received repairs and alterations designed to make her a more effective gunnery training ship. The alterations included the addition of 5-inch/38-caliber guns in single mounts with gun shields similar to those fitted on the more modern types of destroyers then in service. She also lost her prewar colors, being repainted in overall measure one camouflage—dark gray with pale gray tops. With war paint thus donned, Utah sailed for Hawaiian waters on 14 September, after visits to Port Townsend, Wash., and San Francisco and San Pedro, Calif. She arrived at Pearl Harbor soon thereafter and carried out anti-aircraft training and target duties through the late autumn.

Utah completed an advanced anti-aircraft gunnery cruise in Hawaiian waters shortly before she returned to Pearl Harbor in early December 1941, mooring off Ford Island in berth F-11. On the morning of 7 December 1941, the senior officer on board—the captain and executive officer were ashore on leave—was Lt. Comdr. Solomon S. Isquith, the engineer officer.

Shortly before 0800, men topside noted three planes, taken for American planes on maneuvers, heading in a northerly direction from the harbor entrance. They made a low dive at the southern end of Ford Island where the seaplane hangars were situated and began dropping bombs.

The attack on the fleet at Pearl Harbor lasted a little under two hours, but for Utah, it was over in a few minutes. At 0801, soon after sailors had begun raising the colors at the ship's fantail, the erstwhile battleship took a torpedo hit forward, and immediately started to list to port.

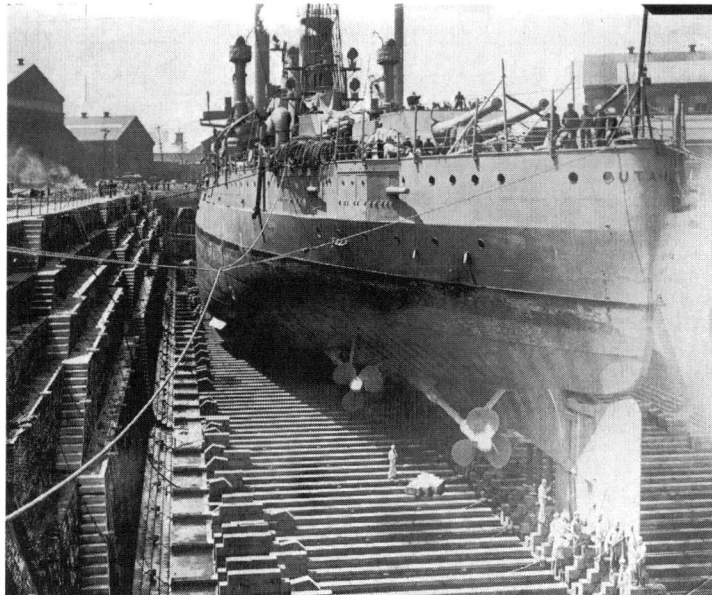

As the ship began to roll ponderously over on her beam ends, 6-by-12-inch timbers—placed on the decks to cushion them against the impact of the bombs used during the ship's latest stint as a mobile target—began to shift, hampering the efforts of the crew to abandon ship. Below, men headed topside while they could. One, however, Chief Watertender Peter Tomich, remained below, making sure that the boilers were secured and that all men had gotten out of the engineering spaces. Another man, Fireman John B. Vaessen, USNR, remained at his post in the dynamo room, making sure that the ship had enough power to keep her lights going as long as possible.

Comdr. Isquith made an inspection to make sure men were out and nearly became trapped himself. As the ship began to turn over, he found an escape hatch blocked. While he was attempting to escape through a porthole, a table upon which he was standing—impelled by the ever-increasing list of the ship—slipped out from beneath him. Fortunately, a man outside grabbed Isquith's arm and pulled him through at the last instant.

At 0812, the mooring lines snapped, and Utah rolled over on her beam ends; her survivors struck out for shore, some taking shelter on the mooring quays since Japanese strafers were active.

Shortly after most of the men had reached shore, Comdr. Isquith, and others, heard a knocking from within the overturned ship's hull. Although Japanese planes were still strafing the area, Isquith called for volunteers to return to the hull and investigate the tapping. Obtaining a cutting torch from the nearby Raleigh (CL-7)—herself fighting for survival after taking an early torpedo hit—the men went to work.

As a result of the persistence shown by Machinist S. A. Szymanski, Chief Machinist's Mate Terrance MacSelwiney, USNR, and two others whose names were unrecorded, 10 men clambered from a would-be tomb. The last man out was Fireman Vaessen, who had made his way to the bottom of the ship when she capsized, bearing a flashlight and wrench.

Utah was declared "in ordinary" on 29 December 1941 and was placed under the control of the Pearl Harbor Base Force. Partially righted to clear an adjacent berth, she was then declared "out of commission, not in service," on 5 September 1944. Utah's name was struck from the Navy list on 13 November 1944. Her partially submerged hulk still remains, rusting, at Pearl Harbor with an unknown number of men trapped inside.

Of Utah's complement, 30 officers and 431 enlisted men survived the ship's loss; 6 officers and 52 men died, four of the latter being recovered and interred ashore. Chief Watertender Tomich received the Medal of Honor posthumously for his selfless act in ensuring the safety of others.

Utah (AG-16) received one battle star for her World War II service.

BB-32 • USS WYOMING

BB- 32: dp. 27,243 (f.), l. 662'0"; b. 93'2Y2", dr. 28'6"
(mean): s. 21.22 k. (tl.); cpl. 1,063; a. 12 12", 21 5", 2 3", 2 21"
tt. ; cl. Wyoming)

Third Wyoming (Battleship No. 32) was laid down
on February 9, 1910, at Philadelphia, PA, by William
Cramp and Sons; launched on May 25, 1911; sponsored
by Miss Dorothy Eunice Knight, the daughter of former
Chief Justice Jesse Knight of the Wyoming Supreme
Court; and commissioned at the Philadelphia Navy
Yard on September 25, 1912, Capt. Frederick L. Chapin
in command.

Wyoming departed Philadelphia on October 6
and completed the fitting-out process at the New York
Navy Yard, Brooklyn, N.Y., before she joined the Fleet
in Hampton Roads, VA. Reaching the Tidewater area
on December 30, 1912, she became the flagship of Rear
Admiral Charles J. Badger, Commander, United States
Atlantic Fleet, soon thereafter. Sailing on January 6, 1913,

the new battleship visited the soon-to-be-completed
Panama Canal and then conducted winter fleet
maneuvers off Cuba before she returned to Chesapeake
Bay on March 4.

After gunnery practice off the Virginia capes on the
southern drill grounds, Wyoming underwent repairs
and alterations at the New York Navy Yard between
April 18 and May 7. She then participated in war games
off Block Island between the 7th and 24th Maya period
of activity broken by repairs to her machinery, carried
out at Newport, R.I., between the 9th and 19th May
before she underwent more repairs at Newport. She then
visited New York City from 28 to May 31 for the festivities
surrounding the dedication of the monument honoring
the battleship Maine, which was destroyed in Havana
harbor on February 15, 1898.

Shifting to Annapolis, Md., on June 4, Wyoming
embarked a contingent of Naval Academy midshipmen
and took the young officers-to-be on a summer cruise

off the coast of New England that lasted into late August. Disembarking the "middies" at Annapolis on 24 and 25 August, Wyoming then conducted torpedo and target practices in the southern drill grounds, out of Hampton Roads, into the late autumn. She was docked at New York for repairs between September 16 and October 2 and then ran a full-power trial as she headed south to Norfolk to resume exercises off the Virginia capes before sailing for Europe on October 26.

Reaching Valetta, Malta, on November 8, the dreadnought-type battleship visited Naples, Italy, and Villefranche, France, during the course of her Mediterranean cruise. The battleship then left French waters astern on the last day of November and reached New York on December 15.

Wyoming then underwent voyage repairs at the New York Navy Yard, remaining there through the end of 1913. Getting underway on January 6, 1914, the battleship reached Hampton Roads on the morrow and spent the next three days coaling to prepare for the annual fleet exercises in the warmer Caribbean climes.

Wyoming exercised with the Fleet out of Guantanamo Bay and Guacanayabo Bay, Cuba, between January 26 and March 15 before setting her course northward for Cape Henry, VA. She then ranged with the Fleet from the southern drill grounds, off the Virginia capes, to Tangier Sound for gunnery drills and practices. She remained engaged in that routine until April 3, when she headed for the New York Navy Yard and an overhaul.

After that period of repairs, which lasted from April 4 to May 9, Wyoming subsequently embarked on a draft of men for transport to the Fleet, departed Hampton Roads on May 13, and headed for Mexican waters. She reached Veracruz on May 18, less than a month after American sailors and marines had occupied that Mexican port.

Wyoming remained at Vera Cruz over the months that ensued, into the late autumn of 1914, before she returned northward. After conducting exercises off the Virginia capes en route, she was put into the New York Navy Yard on October 6 and then underwent repairs and alterations that lasted until January 17, 1915.

Shifting down the coast upon completion of that yard period, Wyoming left Hampton Roads in her wake on January 21 for the annual exercises in Cuban waters and in the Caribbean. Returning to the Tidewater area on April 7, the battleship carried out tactical exercises and maneuvers along the eastern seaboard primarily off Block Island and the southern drill grounds into the late autumn, when she again entered the New York Navy Yard for an overhaul.

After repairs lasting from December 20, 1916, to January 6, 1916, Wyoming got underway on the latter day, bound for war games in the southern drill grounds. She subsequently headed farther south, reaching Culebra, Puerto Rico, on January 16. After visiting Port-au-Prince, Haiti, on January 27, Wyoming put into Guantanamo Bay on the 28th and then operated in Cuban waters off Guantanamo and Guacanayabo Bays and the port of Manzanillo until April 10, when she sailed for New York.

Wyoming remained in the New York Navy Yard from April 16 to June 26, undergoing repairs; she then operated off the New England coast, out of Newport, and off the Virginia Capes through the remainder of 1916. Departing New York on January 9, 1917, Wyoming then conducted routine maneuvers in the Guantanamo Bay region through mid-March. She departed the Caribbean on March 27 and was off Yorktown, VA., when the United States entered World War I on April 6, 1917.

Over the months that ensued, Wyoming served in the Chesapeake Bay region as an engineering ship until November 13, 1917. On that day, Rear Admiral Hugh Rodman broke his flag in New York (Battleship No. 34) as Commander, Battleship Division 9. After preparations for "distant service," Wyoming, New York, Delaware (Battleship No. 28), and Florida (Battleship No. 30) sailed for the British Isles on November 25 and reached Scapa Flow, - Orkney Islands, on December 7, 1917. Although retaining their American designation as Battleship Division 9, those four dreadnoughts became the 6th Battle Squadron of the British Grand Fleet upon arrival in British waters.

Wyoming carried out maneuvers and tactical exercises with the units of the British Grand Fleet until

February 6 1918. On that day, she got underway with the other ships of the 6th Battle Squadron and eight British destroyers to guard a convoy routed to Stavanger, Norway. En route, Wyoming dodged torpedo wakes off Stavanger on February 8 but reached Scapa Flow safely two days later. In the following months, Wyoming continued to patrol off the British Isles, guarding the coastwise sea lanes against the danger posed by the still-powerful German High Seas Fleet.

Between June 30 and July 2 1918, Wyoming operated with the 6th Battle Squadron and a division of British destroyers, guarding Allied minelayers as they planted the North Sea Mine Barrage. Later, Wyoming returned to the Firth of Forth, where she was inspected by the King of England, His Majesty George V, along with other units of the Grand Fleet.

Although American and German capital ships never met in combat on the high seas, they nevertheless made rendezvous. On November 21, 1918, 10 days after the armistice ended World War I, Wyoming, New York, Texas (Battleship No. 35), and Arkansas (Battleship No. 33) joined the Grand Fleet as it escorted the German High Seas Fleet into the Firth of Forth to be interned following the cessation of hostilities.

Later, Wyoming, hoisting the flag of Rear Admiral William S. Sims, Commander, Battleship Division 9, sailed on December 12, 1918, from Portland, England, bound for France. The following morning, she and other battleships rendezvoused with George Washington (Id. No. 3018) off Brest, France. Embarked in the transport was the President of the United States, Woodrow Wilson, en route to the Paris Peace Conference.

After serving in the honor escort for the President and his party, Wyoming returned Admiral Sims to Plymouth, England, along with the newly appointed ambassador to Great Britain. Debarking her distinguished passengers on December 14, the battleship loaded 381 bags of mail and, within a few hours, sailed for the United States. Reaching New York City on Christmas Day 1918, she remained there through New Year's Day 1919. On January 13 1919, she became the flagship of Battleship Division 7, 3d Squadron, and broke the flag of Rear Admiral Robert E. Coontz.

Wyoming departed New York on February 1 and, following winter maneuvers in Cuban waters, returned north, reaching New York on April 14. However, she stood out to sea soon thereafter, getting underway on May 12 to serve as a link in the chain of ships stretching across the Atlantic to guide the N.C. boats on their flight across that ocean. After completing her duty as a plane guard and meteorological station, Wyoming returned to Hampton Roads on the last day of May.

Later, embarking midshipmen and taking them on their southern cruise in the Chesapeake Bay-Virginia capes area, Wyoming entered the Norfolk Navy Yard on July 1 to prepare for service in the Pacific. On that day, she became a unit of the newly designated Pacific Fleet, assigned the duty as flagship for Battleship Division 6, Squadron 4. On the morning of July 19, the Fleet, led by flagship New Mexico (Battleship No. 40), got underway for the Pacific. Transiting the Panama Canal soon thereafter, the Fleet reached San Diego, Calif., on August 6.

Shifting to San Pedro, Calif., three days later, Wyoming operated out of that port into the autumn. After an overhaul at the Puget Sound Navy Yard, Bremerton, Wash., from September 15, 1919, to April 19, 1920, Wyoming returned to her base at San Pedro on May 4. Over the next few months, the battleship exercised off

the southern California coast. During that time, she was reclassified BB-32 on July 17, 1920.

Departing San Diego on the last day of August 1920, Wyoming sailed for Hawaiian waters and conducted exercises and maneuvers there through September. Returning to San Diego on October 3, Wpoming subsequently conducted tactical evolutions off the western seaboard, ranging north to Seattle. Departing San Francisco, Calif., on January 6, 1921, Wyoming, over the ensuing weeks, conducted further drills, exercises, and maneuvers reaching from Panama Bay to Valparaiso, Chile, and was reviewed by the President of Chile on February 3. Returning north via Panama Bay and San Pedro, Wyoming arrived at the Puget Sound Navy Yard on March 18 and remained there into the summer.

Upon completion of repairs, Wyoming headed south and, on August 2, reached Balboa, Canal Zone, where she embarked Rear Admiral Hugh Rodman and members of the commission to Peru for transportation to New York City. Reaching her destination on August 19, she disembarked her passengers and, that afternoon broke the flag of Admiral Hilary P. Jones, the Commander in Chief United States Atlantic Fleet.

Over the next 41 months, Wyoming operated primarily in the Atlantic, off the eastern seaboard of the United States, participating in Atlantic Fleet exercises ranging from the coast of New England to the Virginia Capes. She took part in the routine winter maneuvers of the Fleet in the Caribbean and Cuban waters, serving at various times as flagship for Vice Admiral John D. McDonald, Commander of Battleship Force, and, later Commander, Scouting Fleet, and his successors, Vice Admiral Newton A. McCully and Vice Admiral Josiah S. McKean. During that time, the ship received routine repairs and alterations at the New York Navy Yard and conducted a midshipman's training cruise in the summer of 1924, cruising to Torbay, England, Rotterdam, Holland, Gibraltar, and the Azores.

Departing New York on January 26, 1925, the battleship conducted battle practice in Cuban waters, out of Guantanamo Bay, and then transited the Panama Canal on February 14 to join the Battle Fleet for exercises along the coast of California. Wyoming next sailed for Hawaiian waters and operated in those climes from late April to early June. After a visit to San Diego from 18 to June 22, the battleship returned to the east coast via the Panama Canal and arrived back at New York City on July 17 to resume operations off the coast of New England. Following those training evolutions with a cruise to Cuba and Haiti, Wyoming underwent an overhaul at the New York Navy Yard from November 23, 1925, to January 26, 1926. During her yard period, Comdr. William F. Halsey, Jr., reported on board as the battleship's executive officer. The future fleet admiral served in Wyoming until January 4, 1927.

Wyoming subsequently took part in the Fleet's annual winter maneuvers in the Caribbean and then returned northward, reaching Annapolis on May 29 to embark midshipmen for their summer training cruise. After touching at Newport, R.I., Marblehead, Mass.; Portland, Maine; Charleston, S.C.; and Guantanamo Bay, Wyoming, returned to Annapolis on August 27, disembarking the officers-to-be upon arrival. The ship was then put into the Philadelphia Navy Yard for modernization.

Converted from a coal burner to an oil burner, Wyoming also received new turbine blisters for added underwater protection against torpedoes and other alterations. Completing the overhaul on November 2, 1927, and heading south for Norfolk, Wyoming then underwent a post-modernization shakedown cruise to Cuba and the Virgin Islands before returning to Philadelphia on December 7. Two days later, she hoisted the flag of Commander, Scouting Fleet, Vice Admiral Ashley H. Robertson.

Over the next few years, Wyoming operated out of Norfolk, New York, and Boston, making training cruises for the Naval Reserve Officers' Training Corps (NROTC) units hailing from Yale, Harvard, Georgia Tech, and Northwestern. That duty took her from the Gulf of Mexico to Nova Scotia and into the Caribbean, as well as to the Azores. During the course of that duty, she departed Hampton Roads on November 12, 1928, and, on the night of 13 and 14 November, picked up eight survivors of the sunken British merchant steamship Vestris and landed them at Norfolk the following day, November 15.

Relieved as flagship of the Scouting Force on September 19, 1930, Wyoming then became the flagship of Rear Admiral Wat T. Cluverius, Commander, Battleship Division 2, and performed that duty until November 4. After hoisting the flag of Rear Admiral H. H. Christy, Commander, Training Squadron, Scouting Fleet, the battleship conducted a training cruise into the Gulf of Mexico, during which she visited New Orleans.

Returning north after that cruise, Wyoming was placed in reduced commission at the Philadelphia Navy Yard on January 1, 1931, to prepare for demilitarization and conversion to a training ship in accordance with the 1930 London Treaty for the limitation and reduction of naval armaments. During that process, Wyoming lost her blisters, side armor, and the removal of guns and turret machinery from three of her six main battery turrets. On May 21, 1931, W~oming was relieved of her duties as flagship for the Scouting Force by Augusta (CA-31) and by Arkansas (B.B. 33) as flagship of the Training Squadron.

Wyoming subsequently visited Annapolis upon the completion of her demilitarization and, between May 29 and June 5, 1931, embarked Naval Academy midshipmen for a cruise to European waters. Sailing on June 5, the ship was in the mid-Atlantic ten days later when she went to the aid of the foundering ice-cutting submarine Nautilus, commanded by the famed British Arctic explorer Sir Hubert Wilkins. Wyoming took the disabled submersible in tow and took her to Queenstown, Northern Ireland. Later in the course of the cruise, the former battleship visited Copenhagen, Denmark; Greenock, Scotland; Cadiz, Spain; and Gibraltar before she returned to Hampton Roads on August 13. During her cruise, she was redesignated from a battleship, BB-32, to a miscellaneous auxiliary, AG-17, on July 1, 1931.

Over the next four years, Wyoming continued summer practice cruises for Naval Academy midshipmen and training cruises for NROTC midshipmen with units from various universities. Her service took her throughout the Caribbean and the Gulf of Mexico, as well as to northern European ports and into the Mediterranean.

However, there were new jobs for the old campaigner. On January 18, 1935, she embarked men of the 2nd Battalion, 4th Marine Regiment, at Norfolk for the winter patrol landing assault practices at Puerto Rico and the Panama Canal Zone. In almost every succeeding year, Wyoming took part in amphibious assault exercises as the elements of the Fleet Marine Force and Navy developed tactics for use in possible conflicts of the future.

Departing Norfolk on January 5, 1937, Wyoming transited the Panama Canal, headed for San Diego soon thereafter, and spent the following weeks engaged in assault landing exercises and gunnery drills at San Clemente Island, off the coast of California. On February 18, 1937, during the culminating phase of a multi-faceted (land, sea, and air) exercise, a shrapnel shell exploded prematurely as it was being rammed into one of the ship's 5-inch broadside guns. Six marines were killed, and 11 were wounded. Immediately after the explosion, Wyoming sped to San Pedro, where she transferred the wounded marines to the hospital ship Relief (AH-1).

Completing her slate of exercises and war games off the California coast on March 3, Wyoming stood out of Los Angeles harbor on that day and headed back to the east coast. Returning to Norfolk on the 23rd of the same month, the ship served as temporary flagship for Rear Admiral Wilson Brown, Commander, Training Squadron, from April 15 to June 3, during the preparations for the upcoming Naval Academy practice cruise. Putting to sea on June 4 from Hampton Roads, Wyoming, she reached Kiel, Germany, on June 21, 1937, where she was visited by officers from the ill-fated German "pocket battleship" Admiral Graf Spee. Her embarked midshipmen subsequently toured Berlin before Wyoming sailed for home on June 29, touching at Torbay, England, and Funchal, Madeira, before returning to Norfolk on August 3.

After local exercises, Wyoming disembarked her midshipmen at Annapolis on August 26. For the next few months, Wyoming continued in her role as training ship first for Naval Reserve units and then for Merchant Marine Reserve units, ranging from Boston to the Virgin

Islands and from New York to Cuba, respectively, before she underwent an overhaul at the Norfolk Navy Yard between October 16, 1937, and January 14, 1938.

For the next three years, Wyoming continued her operations out of Norfolk, Boston, and New York, visiting Cuban waters, as well as Puerto Rico and New Orleans. In addition, she conducted a Naval Academy midshipman's practice cruise to European waters in 1938, visiting Le Havre, France, Copenhagen, and Portsmouth, England. Ultimately, on January 2, 941, Wyoming became the flagship for Rear Admiral Randall Jacobs, Commander, Training, Patrol Force, and continued in her training ship duties into the autumn months.

In November 1941, Wyoming embarked on yet another phase of her career that was on a gunnery training ship. She departed Norfolk on November 25, 1941, for gunnery training runs out of Newport, R.I., and was off Platt's Bank when the Japanese attacked Pearl Harbor, Territory of Hawaii, on December 7, 1941.

Putting into Norfolk on January 28, 1942, Wyoming sailed out into the lower reaches of Chesapeake Bay on February 5 to begin a countless chain of gunnery training drills in that area that would carry her through World War II. So familiar was her appearance in the area that Wyoming earned the nickname of the "Chesapeake Raider." Assigned to the Operational Training Command, United States Atlantic Fleet, the former dreadnought battleship provided the platform on which thousands of gunners trained in guns ranging from 5-inch to .50-caliber.

Refitted at Norfolk between January 12 and April 3, 1944, Wyoming took on a different silhouette upon emerging from that yard period; the rest of her 12-inch turrets were removed and replaced with twin-mount 5-inch guns, in addition, newer models of fire control

radars were installed. She resumed her gunnery training activities on April 10, 1944, operating in the Chesapeake Bay region. The extent of her operations can be seen from a random sampling of figures; in a single month, November 1944, Wyoming trained 133 officers and 1,329 men in antiaircraft gunnery. During that month, she fired 3,033 5-inch shells, 849 3-inch, 10,076 40-millimeter, 32,231 20-millimeter, and 66,270 .30-caliber and 360 1.1-inch ammunition. She claimed the distinction of firing off more ammunition than any other ship in the Fleet, training an estimated 35,000 gunners on seven different types of guns.

On June 30, 1945, Wyoming completed her career as "Chesapeake Raider" when she departed Norfolk for the New York Navy Yard and alterations. Leaving the yard on July 13, 1945, she entered Casco Bay soon thereafter, reporting for duty to Vice Admiral Willis A. Lee, Commander, Composite Task Force 69. She fired her first experimental gunnery practice At towed sleeves, drone aircraft, and radio-controlled targets, as the largest operating unit of the force established to study methods and tactics for dealing with the Japanese kamikaze. Subsequently, Composite Task Force 69 became the Operational Development Force, United States Fleet, on August 31, 1945. Upon the death of Admiral Lee, the reins of command passed to Rear Admiral R. P. Briscoe.

Even after the broadening of the scope of the work of the force to cover all the operational testing of new devices of fire control, Wyoming remained the backbone of the unit through 1946. On July 11, 1947, Wyoming entered the Norfolk Naval Shipyard and was decommissioned on August 1, 1947. Her men and materiel were then transferred to Mississippi (AG-128) (ex-BB41)

Wyoming's name was struck from the Navy list on September 16, 1947, and her hulk was sold for scrapping on October 30, 1947. She was then delivered to her purchaser, Lipsett, Inc., of New York City, on December 5, 1947.

BB-33 • USS ARKANSAS

(BB– No. 33: dp. 27,243; l. 562'; b. 93'1 1/2"; dr. 28'6"; s. 21.05 k.; cpl. 1,036; a. 12 12", 215", 2 21" tt.; cl. Wyoming)

The third Arkansas (Battleship No. 33) was laid down on 25 January 1910 at Camden, N.J., by the New York Shipbuilding Co.; launched on 14 January 1911; sponsored by Miss Nancy Louise Macon; and commissioned at the Philadelphia Navy Yard on 17 September 1912, Capt. Roy C. Smith in command.

The new battleship took part in a fleet review by President William H. Taft in the Hudson River off New York City on 14 October and received a visit from the Chief Executive that day. She then transported President Taft to the Panama Canal Zone for an inspection of the unfinished isthmian waterway. After putting the inspection party ashore, Arkansas sailed to Cuban waters for shakedown training. She then returned to the Canal Zone on 26 December to carry President Taft to Key West, Fla.

Following this assignment, Arkansas joined the Atlantic Fleet for maneuvers along the east coast. The battleship began her first overseas cruise in late October 1913 and visited several ports in the Mediterranean. At Naples, Italy, on 11 November 1913, the ship celebrated the birthday of the King of Italy.

Earlier in October 1913, a coup in Mexico had brought to power a dictator, Victoriano Huerta. The way in which Huerta had come to power, however, proved contrary to the idealism of President Woodrow Wilson, who insisted on a representative government rather than a dictatorial one south of the American-Mexican border. Mexico had been in turmoil for several years, and the United States Navy maintained a force of ships in those waters, ready to protect American lives.

In a situation where tension exists between two powers, incidents are bound to occur. One such occurred at Tampico in the spring of 1914, and although the misunderstanding was quickly cleared up locally, the prevailing state of tension produced an explosive situation. Learning that a shipment of arms for Huerta was due to arrive at Veracruz, President Wilson ordered the Navy to prevent the landing of the guns by seizing the customs house at that port.

While a naval force under Rear Admiral Henry T. Mayo was already present in Mexican waters, the President directed that the Atlantic Fleet, under Rear Admiral Charles J. Badger, proceed to Veracruz. Arkansas participated in the landings at Veracruz, contributing a battalion of four companies of bluejackets, a total of 17 officers, and 313 enlisted men under the command of Lt. Comdr. Arthur B. Keating. Among the junior officers was Lt. (jg.) Jonas H. Ingram, who would be awarded the Medal of Honor for heroism at Veracruz, as would Lt. John Grady, who commanded the artillery of the 2nd Seaman Regiment.

Landing on 22 April, Arkansas's men took part in the slow, methodical street fighting that eventually secured the city. Two Arkansas sailors, Ordinary Seamen Louis O. Fried and William L. L. Watson died of their wounds on 22 April. Arkansas's Battalion returned to the ship on 30 April, and the ship remained in Mexican waters through the summer before setting a course on 30 September to return to the east coast. During her stay at Veracruz Z, she received calls from Capt. Franz von Papen, the German military attache to the United States and Mexico, and Rear Admiral Sir Christopher Cradock, on 10 and 30 May 1914, respectively.

The battleship reached Hampton Roads, Va., on 7 October, and after a week of exercises, Arkansas sailed to the New York Navy Yard for repairs and alterations. She then returned to the Virginia capes area for maneuvers on the Southern Drill Grounds. On 12 December, Arkansas returned to the New York Navy Yard for further repairs.

She was underway again on 16 January 1915 and returned to the Southern Drill Grounds for exercises there from 19 to 21 January. Upon completion of these, Arkansas sailed to Guantanamo Bay, Cuba, for fleet exercises. Returning to Hampton Roads on 7 April, the battleship began another training period in the Southern Drill Grounds. On 23 April, she headed to the New York Navy Yard for a two-month repair period. Arkansas then left New York on 25 June bound for Newport, R.I. She conducted torpedo practice and tactical maneuvers in Narragansett Bay through late August.

Returning to Hampton Roads on 27 August, the battleship engaged in maneuvers in the Norfolk area through 4 October, then sailed once again to Newport. There, Arkansas carried out strategic exercises from 5 to 14 October. On 15 October, the battleship arrived at the New York Navy Yard for drydocking. Underway on 8 November, she returned to Hampton Roads. After a period of routine operations, Arkansas went back to Brooklyn for repairs on 19 October. The ship sailed on 5 January 1916 for Hampton Roads. Pausing there only briefly, Arkansas pushed on to the Caribbean for winter maneuvers.

She visited the West Indies and Guantanamo Bay before returning to the United States on 12 March for torpedo practice off Mobile Bay. The battleship then steamed back to Guantanamo Bay on 20 March and remained there until mid-April. On 15 April, the battleship was once again at the New York Navy Yard for overhaul.

On 6 April 1917, the United States entered World War I on the side of the Allied and Associated Powers. The declaration of war found Arkansas attached to Battleship Division 7 and patrolling the York River in Virginia. For the next 14 months, Arkansas carried out patrol duty along the East Coast and trained gun crews for duty on armed merchantmen.

In July 1918, Arkansas received orders to proceed to Rosyth, Scotland, to relieve Delaware (Battleship No. 28).

Arkansas sailed on July 14. On the eve of her arrival in Scotland, the battleship opened fire on what was believed to be the periscope wake of a German U-boat. Her escorting destroyers dropped depth charges but scored no hits. Arkansas then proceeded without incident and dropped anchor at Rosyth on 28 July.

Throughout the remaining three and one-half months of the war, Arkansas and the other American battleships in Rosyth operated as part of the British Grand Fleet as the 6th Battle Squadron.

The armistice ending World War I became effective on 11 November. The 6th Battle Squadron and other Royal Navy units sailed to a point some 40 miles east of May Island at the entrance of the Firth of Forth. Arkansas was present at the internment of the German High Seas Fleet in the Firth of Forth on 21 November 1918.

The American battleships were detached from the British Grand Fleet on 1 December. From the Firth of Forth, Arkansas sailed to Portland, England, thence out to sea to meet the transport George Washington, with President Wilson on board. Arkansas, along with other American battleships, escorted the President's ship into Brest, France, on 13 December 1918. From that French port, Arkansas sailed to New York City, where she arrived on 26 December to a tumultuous welcome. Secretary of the Navy Josephus Daniels reviewed the assembled battleship fleet from the yacht Mayflower.

Following an overhaul at the Norfolk Navy Yard, Arkansas joined the fleet in Cuban waters for winter maneuvers. Soon thereafter, the battleship got underway to cross the Atlantic. On 12 May 1919, she reached Plymouth, England; thence, she headed back out in the Atlantic to take weather observations on 19 May and act as a reference vessel for the flight of the Navy Curtiss (NC) flying boats from Trepassey Bay, Newfoundland, to Europe.

Her role in that venture completed, Arkansas proceeded thence to Brest, where she embarked Admiral William S. Benson, the Chief of Naval Operations, and his wife, on 10 June, upon the admiral's return from the Peace Conference in Paris, before departing for New York. She arrived on 20 June 1919.

Arkansas sailed from Hampton Roads on 19 July 1919 and was assigned to the Pacific Fleet. Proceeding via the Panama Canal, the battleship steamed to San Francisco, where, on 6 September 1919, she embarked Secretary of the Navy and Mrs. Josephus Daniels. Disembarking the Secretary and his wife at Blakely Harbor, Wash., on the 12th, Arkansas was reviewed by President Wilson, and on the 13th, the Chief Executive embarked in the famed Oregon (Battleship No. 3). On 19 September 1919, Arkansas entered the Puget Sound Navy Yard for a general overhaul. Resuming her operations with the fleet in May 1920, Arkansas operated off the California coast. On 17 July 1920, Arkansas received the designation BB-33 as the ships of the fleet received alphanumeric designations. That September, she cruised to Hawaii for the first time. Early in 1921, the battleship visited Valparaiso, Chile, manning the rail in honor of the Chilean President.

Arkansas's peacetime routine consisted of an annual cycle interspersed with periods of upkeep or overhaul. The battleship's schedule also included competitions in gunnery and engineering and an annual fleet problem. Becoming flagship for the Commander, Battleship Force, Atlantic Fleet, in the summer of 1921, Arkansas began operations off the east coast that August.

For a number of years, Arkansas was detailed to take midshipmen from the Naval Academy on their summer cruises. In 1923, the battleship steamed to Europe, visiting Copenhagen, Denmark (where she was visited by the King of Denmark on 2 July 1923), Lisbon, Portugal, and Gibraltar. Arkansas conducted another midshipman training cruise to European waters the following year, 1924. In 1925, the cruise was to the west coast of the United

States. During this time, on 30 June 1925, Arkansas arrived at Santa Barbara, California, in the wake of an earthquake. The battleship, along with McCawley (DD-276) and Eagle 34 (PE-34), landed a patrol of bluejackets for policing Santa Barbara and established a temporary radio station ashore for the transmission of messages.

Upon completion of the 1925 midshipman cruise, Arkansas entered the Philadelphia Navy Yard for modernization. Her coal-burning boilers were replaced with oil-fired ones. Additional deck armor was installed, a single stack was substituted for the original pair, and the after-cage mast was replaced by a low tripod. Arkansas left the yard in November 1926 and, after a shakedown cruise along the eastern seaboard and to Cuban waters, returned to Philadelphia to run acceptance trials. Resuming her duty with the fleet soon thereafter, she operated from Maine to the Caribbean; on 5 September 1927, she was present at ceremonies unveiling a memorial tablet honoring the French soldiers and sailors who died during the campaign at Yorktown in 1781.

In May 1928, Arkansas again embarked midshipmen for their practice cruise along the eastern seaboard and down into Cuban waters. During the first part of 1929, she operated near the Canal Zone and in the Caribbean, returning in May 1929 to the New York Navy Yard for overhaul. After embarking midshipmen at Annapolis, Arkansas, carried out her 1929 practice

cruise to Mediterranean and English waters, returning in August to operate with the Scouting Fleet off the east coast.

In 1930 and 1931, Arkansas was again detailed to carry out midshipmen's practice cruises; in the former year, she visited Cherbourg, France; Kiel, Germany; Oslo, Norway; and Edinburgh, Scotland; in the latter, her itinerary included Copenhagen, Denmark; Greenock, Scotland; and Cadiz, Spain, as well as Gibraltar. In September 1931, the ship visited Halifax, Nova Scotia. In October, Arkansas participated in the Yorktown Sesquicentennial celebrations, embarking President Herbert Hoover and his party on 17 October and taking them to the exposition. She later transported the Chief Executive and his party back to Annapolis on 19 and 20 October. Upon her return, the battleship entered the Philadelphia Navy Yard, where she remained until January 1932.

Upon leaving the navy yard, Arkansas sailed for the west coast, calling at New Orleans, La., en route to participate in the Mardi Gras celebration. Assigned duty as flagship of the Training Squadron, Atlantic Fleet, Arkansas, operated continuously on the west coast of the United States into the spring of 1934, at which time she returned to the east coast.

In the summer of 1934, the battleship conducted a midshipman in Plymouth, England; Nice, France; Naples, going to Annapolis in August, where she manned the rail for lt as he passed on board the yacht Nourmalhal and was present for the International Yacht Race. Arkansas' cutter defeated the cutter from the British light cruiser HMS Dragon for the Battenberg Cup and the City of Newport.

In January 1935, Arkansas transported the 1st Battalion, 5th Marines, to Culebra for a fleet landing exercise, and in June, conducted a midshipman practice cruise to Europe, visiting Edinburgh, Oslo (where King Haakon VII of Norway visited the ship), Copenhagen, Gibraltar, and Funchal on the island of Madeira. After disembarking Naval Academy midshipmen at Annapolis in August 1935, Arkansas proceeded to New York K. There, she embarked reservists from the New York area and conducted a Naval Reserve cruise to Halifax, Nova

Scotia, in September. Upon completion of that duty, she underwent repairs and alterations at the New York Navy Yard that October.

In January 1936, Arkansas participated in Fleet Landing Exercise No. 2 at Culebra and then visited New Orleans for the Mardi Gras festivities before she returned to Norfolk for a navy yard overhaul, which lasted through the spring of 1936. That summer, she carried out a midshipman training cruise to Portsmouth, England; Goteborg, Sweden; and Cherbourg before she returned to Annapolis that August. Steaming thence to Boston, the battleship conducted a Naval Reserve training cruise before putting into the Norfolk Navy Yard for an overhaul that October.

The following year, 1937, saw Arkansas make a midshipman practice cruise to European waters, visiting ports in Germany and England before she returned to the east coast of the United States for local operations out of Norfolk. During the latter part of the year, the ship also ranged from Philadelphia and Boston to St. Thomas, Virgin Islands, and Cuban waters. During 1938 and 1939, the pattern of operations largely remained as it had been in previous years; her duties in the Training Squadron largely con- her to the waters of the eastern seaboard.

The outbreak of war in Europe in September 1939 found Arkansas at Hampton Roads, preparing for a Naval Reserve cruise. She soon got underway and transported seaplane moorings and aviation equipment from the naval air station at Norfolk to Narragansett Bay for the seaplane base that was to be established there. While at Newport, Arkansas, took on board ordnance material for destroyers and brought it back to Hampton Roads.

Arkansas departed Norfolk on 11 January 1940, in company with Texas (BB-35) and New York (BB-34), and proceeded thence to Guantanamo Bay for fleet exercises. She then participated in landing exercises at Culebra that February, returning via St. Thomas and Culebra to Norfolk. Following an overhaul at the Norfolk Navy Yard (18 March to 24 May), Arkansas shifted to the Naval Operating Base (NOB), Norfolk, where she remained until 30 May. Sailing on that day for Annapolis, the battleship, along with Texas and New York, conducted a midshipman training cruise to Panama and Venezuela that summer. Before the year was out, Arkansas would conduct three V-7 Naval Reserve training cruises, these voyages taking her to Guantanamo Bay, the Canal Zone, and Chesapeake Bay.

Over the months that followed, the United States gradually edged toward war in the Atlantic; early the following summer, after the decision to occupy Iceland had been reached, Arkansas

accompanied the initial contingent of marines to that place. That battleship, along with New York and the light cruiser Brooklyn (CL-40), provided the heavy escort for the convoy. Following this assignment, Arkansas sailed to Casco Bay, Maine, and was pres- cruised to Italy and to Gibraltar, returning proceeding thence to Newport, R. I. There when, the Atlantic Charter conferences took place on board Augusta (CA-31) between President Franklin D. Roosevelt and British Prime Minister Winston Churchill. During the conference, the battleship provided accommodations for the Under Secretary of State, Sumner Welles, and Mr. Averell Harriman, from 8 to 14 August 1941.

The outbreak of war with the Japanese attack upon the Pacific Fleet at Pearl Harbor found Arkansas at anchor in Casco Bay, Maine. One week later, on 14 December, she sailed to Hvalfjordur, Iceland. Returning to Boston via Argentia on 24 January 1942, Arkansas spent the month of February carrying out role exercises in Casco Bay in preparation for her role as an escort for troop and cargo transports. On March 6, she arrived at Norfolk to begin the overhaul. Underway on 2 July, Arkansas conducted shakedown in Chesapeake Bay, then proceeded to New York City, where she arrived on 27 July.

The battleship sailed from New York on 6 August, bound for Greenock, Scotland. Two days later, the ships paused at Halifax, Nova Scotia, then continued on through the stormy North Atlantic. The convoy reached Greenock on the 17th, and Arkansas returned to New York on 4 September. She escorted another Greenock-bound convoy across the Atlantic, then arrived back at New York on 20 October. With the Allied invasion of North Africa, American convoys were routed to

Casablanca to support the operations. Departing New York on 3 November, Arkansas covered a troop convoy to Morocco and returned to New York on 11 December for overhaul.

On 2 January 1943, Arkansas sailed to Chesapeake Bay for gunnery drills. She returned to New York on 30 January and began loading supplies for yet another transatlantic trip. The battleship made two runs between Casablanca and New York City from February through April. In early May, Arkansas was drydocked at the New York Navy Yard, emerging from that period of yard work to proceed to Norfolk on 26 May.

Arkansas assumed her new duty as a training ship for midshipmen based at Norfolk. After four months of operations in the Chesapeake Bay, the battleship returned to New York to resume tier role as a convoy escort. On 8 October, the ship sailed for Bangor, Ireland. She was in that port throughout November and got underway to return to New York in December. Arkansas then began a period of repairs on 12 December. Clearing New York for Norfolk two days after Christmas of 1943, Arkansas closed the year in that port.

The battleship sailed on 19 January 1944 with a convoy bound for Ireland. After seeing the convoy safely to its destination, the ship reversed its course across the Atlantic and reached New York on 13 February. Arkansas went to Casco Bay on 28 March for gunnery exercises before she proceeded to Boston on 11 April for repairs.

On 18 April, Arkansas sailed once more for Bangor, Ireland. Upon her arrival, the battleship began a training period to prepare for her new role as a shore bombardment ship. On 3 June, Arkansas sailed for the French coast to support the Allied invasion of Normandy. The ship entered the Baie de la Seine on 6 June and took up a position 4,000 yards off "Omaha" beach. At 0552, Arkansas's guns opened fire. During the day, the

venerable battleship underwent shore battery fire and air attacks; over the ensuing days, she continued her fire support. On the 13th, Arkansas shifted to a position off Grandcamp les Bains.

On 25 June 1944, Arkansas dueled with German shore batteries off Cherbourg, the enemy repeatedly straddling the battleship but never hitting her. Her big guns helped support the Allied attack on that key port and led to the capture of it the following day. Retiring to Weymouth, England, and arriving there at 2220, the battleship shifted to Bangor on 30 June.

Arkansas stood out to sea on 4 July, bound for the Mediterranean. She passed through the Strait of Gibraltar and anchored at Oran, Algeria, on 10 July. On the 18th, she got underway and reached Taranto, Italy, on 21 July. The battleship remained there until 6 August, then shifted to Palermo, Sicily, on the 7th.

On 14 August, Operation "Anvil," the invasion of the southern French coast between Toulon and Cannes, began. Arkansas provided fire support for the initial landings on 15 August and continued her bombardment through 17 August. After stops at Palermo and Oran, Arkansas, set course for the United States. On 14 September, she reached Boston and received repairs and alterations through early November. The yard period was completed on 7 November; Arkansas sailed to Casco Bay for three days of refresher training. On 10 November, Arkansas shaped a course south for the Panama Canal Zone. After transiting the canal on 22 November, Arkansas headed for San Pedro, Calif. On 29 November, the ship was again underway for exercises held off San Diego. She returned to San Pedro on 10 December.

After three more weeks of preparations, Arkansas sailed for Pearl Harbor on January 20, 1945. One day after her arrival there, she sailed for Ulithi, the major fleet staging area in the Carolines, and continued thence to Tinian, where she arrived on 12 February. For two days, the vessel held shore bombardment practice prior to her participation in the assault on Iwo Jima.

At 0600 on 16 February, Arkansas opened fire on Japanese strong points on Iwo Jima as she lay off the island's west coast. The old battlewagon bombarded the island through the 19th century and remained in the fire support area to provide cover during the evening hours. During her time off the embattled island, Arkansas shelled numerous Japanese positions in support of the bitter struggle by the Marines to root out and destroy the stubborn enemy resistance. She cleared the waters off Iwo Jima on 7 March to return to Ulithi. After arriving at that atoll on the 10th, the battleship rearmed, provisioned, and fueled in preparation for her next operation, the invasion of Okinawa.

Getting underway on 21 March, Arkansas began her preliminary shelling of Japanese positions on Okinawa on 25 March, some days ahead of the assault troops, which began wading ashore on 1 April. The Japanese soon began an aerial onslaught, and Arkansas fended off several kamikazes. For 46 days, Arkansas delivered fire support for the invasion of Okinawa. On 14 May, the ship arrived at Apra Harbor, Guam, to await further assignment.

After a month at Apra Harbor, part of which she spent in drydock, Arkansas, she got underway on 12 June for Leyte Gulf. She anchored there on the 16th and remained in Philippine waters until the war drew to a close in August. On the 20th of that month, Arkansas left Leyte to return to Okinawa and reached Buckner Bay on August 23. After a month spent in port, Arkansas embarked approximately 800 troops for transport to the United States as part of the "Magic Carpet" to return American servicemen home as quickly as possible. Sailing on 23 September, Arkansas paused briefly at Pearl Harbor en route and ultimately reached Seattle on 15 October. During the remainder of the year, the battleship made three more trips to Pearl Harbor to shuttle soldiers back to the United States.

During the first months of 1946, Arkansas lay at San Francisco. In late April, the ship got underway for Hawaii. She reached Pearl Harbor on 8 May and stood out of Pearl Harbor on 20 May, bound for Bikini Atoll, earmarked for use as a target for atomic bomb testing in Operation "Crossroads." On 25 July 1946, the venerable battleship was sunk in Test "Baker" at Bikini. Decommissioned on 29 July 1946, Arkansas was struck from the Naval Vessel Register on 15 August 1946.

Arkansas received four battle stars for her World War II service.

BB-34 • USS NEW YORK

(BB-34: dp. 27,000; l. 573'; b. 95'3"; dr. 28'6"; s. 21 k.; cpl. 1,042; a. 10 14", 21 5", 4 21" tt.; cl. New York)

The fifth shipped named New York (BB-34) was laid down on September 11, 1911, by the Brooklyn Navy Yard, New York. She was launched on October 30, 1912, sponsored by Miss Elsie Calder, and commissioned on April 15, 1914, with Captain Thomas S. Rodgers in command.

Ordered south shortly after commissioning, New York served as the flagship for Rear Admiral Frank Fletcher, commanding the fleet occupying and blockading Vera Cruz until the resolution of the crisis with Mexico in July 1914. New York then headed north for fleet operations along the Atlantic coast as war broke out in Europe.

Upon the United States' entry into the war, New York, as flagship with Battleship Division 9 commanded by Rear Admiral Hugh Rodman, sailed to strengthen the British Grand Fleet in the North Sea, arriving at Scapa Flow on December 7, 1917. As a separate squadron

within the Grand Fleet, the American ships joined in blockade and escort missions, deterring the Germans from any major fleet engagements. New York twice encountered U-boats.

During her World War I service, New York hosted royal and high-ranking representatives of the Allies and was present for the surrender of the German High Seas Fleet in the Firth of Forth on November 21, 1918. As a final European mission, New York joined the ships escorting President Woodrow Wilson from an ocean rendezvous to Brest en route to the Versailles Conference.

Returning to a program alternating between individual and fleet exercises with necessary maintenance, New York trained in the Caribbean in spring 1919 and joined the Pacific Fleet at San Diego, her home port for the next 16 years. She trained off Hawaii and the West Coast, occasionally returning to the Atlantic and Caribbean for brief missions or overhauls. In 1937, carrying Admiral Hugh Rodman, the President's personal representative for the coronation of King George VI of England, New York sailed to participate in the Grand Naval Review on May 20, 1937, as the sole U.S. Navy representative.

For much of the following three years, New York trained Naval Academy midshipmen and other prospective officers with cruises to Europe, Canada, and the Caribbean. In mid-1941, she joined the Neutrality Patrol, escorting troops to Iceland in July 1941 and serving as a station ship at Argentia, Newfoundland, protecting the new American base. From America's entry into World War II, New York guarded Atlantic convoys to Iceland and Scotland during the peak of the U-boat menace, successfully bringing convoys to harbor intact.

New York provided crucial gunfire support at Safi during the invasion of North Africa on November 8, 1942. She then stood by at Casablanca and Fedhala before returning home for convoy duty, escorting critically needed men and supplies to North Africa.

She then trained gunners for battleships and destroyer escorts in Chesapeake Bay, rendering this vital service until June 10, 1944. She conducted three training cruises for the Naval Academy, voyaging to Trinidad on each.

New York sailed for the West Coast on November 21, arriving in San Pedro on December 6 for gunnery training in preparation for amphibious operations. She departed San Pedro on January 12, 1945, called at Pearl Harbor, and diverted to Eniwetok to survey screw damage. Despite impaired speed, she joined the Iwo Jima assault force in rehearsals at Saipan and participated in pre-invasion bombardment at Iwo Jima on February 16. During the next three days, she fired more rounds than any other ship present and made a spectacular direct 14"-hit on an enemy ammunition dump.

Leaving Iwo Jima, New York repaired her propellers at Manus and regained speed for the assault on Okinawa, which she reached on March 27 to begin 76 consecutive days of action. She fired pre-invasion and diversionary

bombardments, covered landings, and provided close support to advancing troops. On April 14, a kamikaze grazed her, demolishing her spotting plane. She left Okinawa on June 11 to regun at Pearl Harbor.

New York prepared at Pearl Harbor for the planned invasion of Japan and, after the war's end, transported veterans to the West Coast and replacements to Pearl Harbor. She sailed from Pearl Harbor on September 29 with passengers for New York, arriving on October 19. There, she prepared to serve as a target ship in Operation "Crossroads," the Bikini atomic tests, and sailed on March 4, 1946, for the West Coast. She left San Francisco on May 1 and, after calls in Pearl Harbor and Kwajalein, reached Bikini on June 15. Surviving the surface blast on July 1 and the underwater explosion on July 25, she was decommissioned at Kwajalein on August 29, 1946. Later towed to Pearl Harbor, she was studied for two years, and on July 8, 1948, was towed out to sea and sunk after an eight-hour barrage by ships and planes conducting full-scale battle maneuvers with new weapons.

New York earned three battle stars for her World War II service.

BB-35 • USS TEXAS

(BB– 35: displacement 27,000 (normal); length 673'0"; beam 95'2½" (waterline); draft 29'7" (forward); speed 21.06 knots (trial); complement 954; armament 10 14-inch guns, 21 6-inch guns, 4 3-pounders, 4 21-inch torpedo tubes (submerged); class New York)

The second ship named Texas (Battleship No. 36) was laid down on 17 April 1911 at Newport News, Virginia, by the Newport News Shipbuilding Co., launched on 18 May 1912, sponsored by Miss Claudia Lyon, and commissioned on 12 March 1914 with Captain Albert W. Grant in command.

On 24 March, Texas departed the Norfolk Navy Yard and set a course for New York. She made an overnight stop at Tompkinsville, New York, on the night of the 26th and 27th and entered the New York Navy Yard on the latter day. She spent the next three weeks there undergoing the installation of fire control equipment.

During her stay in New York, President Woodrow Wilson ordered a number of ships of the Atlantic Fleet to Mexican waters in response to tensions created when Mexican Federal troops detained an American boat crew at Tampico. The issue was quickly resolved locally, but fiery Rear Admiral Henry T. Mayo sought further redress by demanding an official disavowal of the act by the Huerta regime and a 21-gun salute to the American flag.

Unfortunately for Mexican-American relations, President Wilson saw in the incident an opportunity to pressure a government he deemed undemocratic. On 20 April, Wilson brought the matter before Congress and sent orders to Rear Admiral Frank Friday Fletcher, commanding the naval force off the Mexican coast, instructing him to land a force at Veracruz and seize the customs house there in retaliation for the "Tampico Incident." That action was carried out on the 21st and 22nd.

Due to the intensity of the situation, Texas put to sea on 13 May, heading directly to operational duty without the usual shakedown cruise and post-shakedown repair period. After a five-day stop at Hampton Roads between 14 and 19 May, she joined Rear Admiral Fletcher's force off Veracruz on the 26th. She remained in Mexican waters for just over two months, supporting American forces ashore. On 8 August, she left Veracruz, set a course for Nipe Bay, Cuba, and then steamed to New York where she entered the Navy Yard on 21 August.

The battleship remained there until 6 September, when she rejoined the Atlantic Fleet and settled into a schedule of normal fleet operations. In October, she returned to the Mexican coast and became the station ship at Tuxpan, a duty that lasted until early November. She bid farewell to Mexico at Tampico on 20 December and set a course for New York, entering the New York Navy Yard on 28 December and undergoing repairs until 16 February 1916.

Upon her return to active duty, Texas resumed a schedule of training operations along the New England coast and off the Virginia Capes, alternated with winter fleet tactical and gunnery drills in the West Indies. This routine lasted just over two years until the crisis over unrestricted submarine warfare catapulted the United States into war with the Central Powers in April 1917.

The declaration of war on 6 April found Texas anchored in the mouth of the York River with other Atlantic Fleet battleships. She stayed in the Virginia Capes-Hampton Roads area until mid-August, conducting exercises and training naval armed guard gun crews for service on merchant ships.

In August, she steamed to New York for repairs, arriving at Base 10 on the 19th and entering the New York Navy Yard soon after. She completed repairs on 26 September and got underway for Port Jefferson that same day. However, during the mid-watch on the 27th, she ran aground on Block Island. Despite efforts over three days, she remained stuck until tugs assisted on the 30th. Hull damage necessitated a return to the yard, preventing her from departing with Division 9 for the British Isles in November.

By December, she had completed repairs and moved south to conduct war games out of the York River. In mid-January 1918, Texas prepared for a voyage across the Atlantic. She departed New York on 30 January, arrived at Scapa Flow in the Orkney Islands off the coast of Scotland on 11 February, and rejoined Division 9, now known as the 6th Battle Squadron of Britain's Grand Fleet.

Texas' service with the Grand Fleet consisted of convoy missions and occasional forays to reinforce the British squadron on blockade duty in the North Sea. The fleet alternated between bases at Scapa Flow and the Firth of Forth in Scotland. Texas began her mission only five days after arriving at Scapa Flow, where she sortied with the entire fleet to reinforce the 4th Battle Squadron in the North Sea. She returned to Scapa Flow the next day and remained until 8 March when she embarked on a convoy escort mission, returning on the 13th. Texas and her division mates entered the Firth of Forth on 12 April but got underway again on the 17th to escort a convoy. The American battleships returned to base on 20 April. Four days later, Texas again set out to sea to support the 2nd Battle Squadron the day after the German High Seas Fleet had sortied from Jade Bay toward the Norwegian coast to threaten an Allied convoy. The Germans returned to their base on the 25th, and the Grand Fleet, including Texas, did likewise the next day.

Texas and her division mates experienced a relatively quiet May in the Firth of Forth. On 9 June, she got

underway with the other warships of the 6th Battle Squadron and headed back to Scapa Flow, arriving the following day. From 30 June to 2 July, Texas and her colleagues escorted American minelayers adding to the North Sea mine barrage. After a two-day return to Scapa Flow, Texas put to sea with the Grand Fleet to conduct two days of tactical exercises and war games. At the conclusion of those drills on 8 July, the fleet entered the Firth of Forth. For the remainder of World War I, Texas and the other battleships of Division 9 continued to operate with the Grand Fleet as the 6th Battle Squadron. With the German Fleet increasingly confined to its bases in the estuaries of the Jade and Ems Rivers, the American and British ships settled into a routine schedule of operations with little hint of combat. This state of affairs lasted until the armistice ended hostilities on 11 November 1918. On the night of 20 and 21 November, Texas accompanied the Grand Fleet to meet the surrendering German Fleet.

The two fleets rendezvoused about 40 miles east of May Island, near the mouth of the Firth of Forth, and proceeded together into the anchorage at Scapa Flow. Afterward, the American contingent moved to Portland, England, arriving there on 4 December.

Eight days later, Texas set sail with Divisions 9 and 6 to meet President Woodrow Wilson, embarked in George Washington, on his way to the Paris Peace Conference. The rendezvous took place around 0730 the following morning, and Texas provided an escort for the President into Brest, France, where the ships arrived at 1230 that afternoon. That evening, Texas and the other American battleships departed Brest for Portland, where they stopped briefly on the 14th before getting underway to return to the United States. The warships arrived off Ambrose Light on Christmas Day 1918 and entered New York on the 26th.

Following an overhaul, Texas resumed duty with the Atlantic Fleet in early 1919. On 9 March, she became the first American battleship to launch an airplane when Lt. Comdr. Edward O. McDonnell flew a British-built Sopwith "Camel" off the warship. That summer, she was reassigned to the Pacific Fleet. On 17 July 1920, she was designated BB-35 as a result of the Navy's adoption of the alphanumeric system of hull designations. Texas served in the Pacific until 1924, when she returned to the east coast for an overhaul and a training cruise to European waters with Naval Academy midshipmen embarked. That fall, she conducted maneuvers as a unit of the Scouting Fleet. In 1925, she entered the Norfolk Navy Yard for a major modernization overhaul, during which her cage masts were replaced with a single tripod foremast, and she received the latest in fire control equipment. Following the overhaul, she resumed duty along the eastern seaboard and continued at that task until late in 1927, when she briefly toured the Pacific between late September and early December.

Near the end of the year, Texas returned to the Atlantic and resumed her normal duties with the Scouting Fleet. In January 1928, she transported President Herbert Hoover to Havana for the Pan-American Conference, then continued via the Panama Canal to the West Coast for maneuvers near Hawaii.

She returned to New York early in 1929 for her annual overhaul, completing it by March. She then began another brief tour of duty in the Pacific. By June, she was back in the Atlantic, resuming normal duties with the Scouting Fleet. In April 1930, Texas took a break from her schedule to escort SS Leviathan into New York. This ship carried the delegation that had represented the United States at the London Naval Conference. In January 1931, she left the New York yard as the flagship of the United States Fleet, heading via the Panama Canal to San Diego, her home port for the next six years. During this time, she served first as the flagship for the entire Fleet, and later, as the flagship for Battleship Division I. She left the Pacific once during this period, in the summer of 1936, for a midshipman training cruise in the Atlantic. After completing this assignment, she immediately rejoined the Battle Force in the Pacific.

In the summer of 1937, Texas was reassigned to the East Coast as the flagship of the Training Detachment, United States Fleet. Late in 1938 or early in 1939, she became the flagship of the newly organized Atlantic Squadron, built around BatDiv 5. Her duties primarily involved training missions, midshipman cruises, naval reserve drills, and training members of the Fleet Marine Force.

When war broke out in Europe in September 1939, Texas began operating on the "neutrality patrol," established to keep the war out of the Western Hemisphere. As the United States moved towards more active support of the Allied cause, Texas started convoying ships carrying Lend-Lease material to Great Britain. On Sunday, December 7, 1941, she was at Casco Bay, Maine, for a rest and relaxation period following three months of watch duty at Argentia, Newfoundland. After ten days at Casco Bay, she returned to Argentia and stayed there until late January 1942, when she escorted a convoy to England. After delivering her charges, she patrolled near Iceland until March, then returned home. For the next six months, she continued convoy-escort missions to various destinations. On one occasion, she escorted Guadalcanal-bound Marines as far as Panama. On another, she screened service troops to Freetown, Sierra Leone, on the West Coast of Africa. More frequently, her voyages were to and from Great Britain, escorting both cargo- and troop-carrying ships.

On October 23, Texas embarked on her first major combat operation with Task Group 34.8, the Northern Attack Group for Operation "Torch," the invasion of North Africa. Her objective was Mehedia near Port Lyautey and the port itself. The ships arrived off the assault beaches on the morning of November 8 and began preparations for the invasion. When the troops went ashore, Texas did not immediately enter action to support them. At that time, amphibious warfare doctrine was still developing, and many did not recognize the value of a pre-landing bombardment. The Army insisted on attempting surprise. Texas finally entered the fray early in the afternoon when the Army requested her to destroy an ammunition dump near Port Lyautey. For the next week, she cruised up and down the Moroccan coast, delivering specific call-fire missions. Unlike in later operations, she expended only 273 rounds of 14-inch and 6 rounds of 6-inch ammunition. During her short stay, some of her crew briefly went ashore to assist in salvaging shipping sunk in the harbor. On November 15, she departed North Africa, returning home in company with Savannah (CL-42), Sangamon (ACV-26), Kennebec (AO-36), four transports, and seven destroyers.

Throughout 1943, Texas continued her role as a convoy escort. With New York as her home port, she made numerous transatlantic voyages to places like Casablanca and Gibraltar, as well as frequent visits to ports in the British Isles. This routine continued into 1944 but ended in April of that year when, at the European end of one such mission, she remained at the Clyde estuary in Scotland, training for the invasion of Normandy. This warm-up period lasted about seven weeks, after which she departed the Clyde, traveled down the Irish Sea, and around the southern coast of England to arrive off the Normandy beaches on the night of June 5-6.

At about 0440 on the morning of June 6, the battleship closed in on the Normandy coast to a point about 12,000 yards offshore near Pointe du Hoc. At 0550, Texas began bombarding the coastal landscape with

her 14-inch salvoes. Her secondary battery targeted another area on the western end of "Omaha" beach, a ravine laced with strong points to defend an exit road. Later, under the control of airborne spotters, she shifted her major-caliber fire inland to interdict enemy reinforcement activities and to destroy batteries and other strong points farther inland.

By noon, she had closed in on the beach to about 3,000 yards to fire upon snipers and machine-gun nests hidden in a defile just off the beach. After completing that mission, she targeted an enemy anti-aircraft battery west of Vierville.

The following morning, her main battery bombarded the enemy-held towns of Surrain and Trevieres to break up German troop concentrations. That evening, she targeted a German mortar battery that had been shelling the beach. Shortly after midnight, German planes attacked the ships offshore, but Texas' anti-aircraft batteries, though they opened fire immediately, failed to score a hit. On the morning of June 8, she fired on Isigny, then a shore battery, and finally Trevieres once more.

After this, she retired to Plymouth to rearm, returning to the French coast on June 11. From then until June 15, she supported the Army in its advance inland. However, by June 15, the troops had advanced beyond the range of her guns, and Texas moved on to another mission.

On the morning of June 26, Texas closed in on the vital port of Cherbourg, joining Arkansas (BB-33) to open fire on various fortifications and batteries surrounding the town. The guns on shore returned fire immediately. Around 1230, they managed to straddle Texas. Despite shell geysers erupting around her, the battleship continued her firing runs. The enemy gunners were both stubborn and skilled. At 1316, a 280-millimeter shell hit her fire control tower, killing the helmsman and wounding

nearly everyone on the navigation bridge. Captain Baker, the commanding officer, miraculously escaped unharmed and quickly cleared the bridge. Despite the damage and casualties, Texas continued to deliver her 14-inch shells. Later, another shell, a 240-millimeter armor-piercing shell, struck the battleship. It crashed through the port bow, entered a compartment below the wardroom, but failed to explode. Throughout the three-hour duel, the Germans straddled and nearly missed Texas over 66 times, but she persisted in her mission until 1600, when she received orders to retire.

Texas underwent repairs at Plymouth, England, then prepared for the invasion of Southern France. On July 16, she departed Belfast Lough for the Mediterranean. After stops at Gibraltar and Oran in Algeria, she rendezvoused with three French destroyers off Bizerte, Tunisia, and set a course for the Riviera coast of France. She arrived off St. Tropez during the night of July 14-15. At 0444, she moved into position for the pre-landing bombardment and, at 0651, opened fire on her first target, a battery of five 155-millimeter guns.

The troops ashore moved inland rapidly against light resistance, so Texas provided fire support for the assault for only two days. She departed the Southern coast of France on the evening of August 16. After a stop at Palermo, Sicily, she left the Mediterranean and headed for New York, arriving there on September 14, 1944.

At New York, Texas underwent an 86-day repair period during which the barrels of her main battery were replaced. After a brief refresher cruise, she departed New York in November and set a course, via the Panama Canal, for the Pacific. She made a stop at Long Beach, California, then continued to Oahu. She spent Christmas at Pearl Harbor and then conducted maneuvers in the Hawaiian Islands for about a month, after which she steamed to Ulithi Atoll. Departing Ulithi on February 10, 1945, she stopped in the Marianas for two days of invasion rehearsals, then set a course for Iwo Jima. She arrived off the target on February 16, three days before the scheduled assault, spending those three days bombarding enemy defenses on Iwo Jima in preparation for the landings. After the troops stormed ashore on February 19, Texas switched roles and began delivering support and call fire. She remained off Iwo Jima for almost a fortnight, helping the Marines subdue a well-dug-in and stubborn Japanese garrison.

Although Iwo Jima was not declared secured until March 16, the USS Texas cleared the area in late February and returned to Ulithi in early March to prepare for the Okinawa operation. She departed Ulithi with Task Force 64, the gunfire support unit, on March 21 and arrived in the Ryukyus on the 25th. Texas did not participate in the occupation of the islands and the roadstead at Kerama Retto, which was carried out on the 26th, but instead moved in on the main objective, beginning the pre-landing bombardment that same day. For the next six days, she delivered 14-inch salvos to prepare the way for the Army and the Marine Corps. Each evening, she retreated from her bombardment position close to the Okinawan shore, only to return the next day and resume her shelling. The enemy ashore, preparing for a defense-in-depth strategy as at Iwo Jima, did not respond. Only their air units responded, sending several kamikaze raids to harass the bombardment group. Texas escaped damage during these small attacks. After six days of aerial and naval bombardment, the ground troops began their assault on April 1, storming ashore against initially light resistance. For almost two months, Texas remained in Okinawan waters, providing gunfire support for the troops ashore and fending off enemy aerial assaults. In this latter mission, she claimed one kamikaze kill on her own and three assists.

Late in May, Texas retired to Leyte in the Philippines and remained there until after Japan's capitulation on August 16. She returned to Okinawa toward the end of August and stayed in the Ryukyus until September 23. On that day, she set a course for the United States with troops embarked. The battleship delivered her passengers to San Pedro, California, on October 16. She celebrated Navy

Day there on October 27 and then resumed her mission of bringing American troops home. She made two round-trip voyages between California and Oahu in November and a third in late December.

On January 21, 1946, the warship departed San Pedro and steamed via the Panama Canal to Norfolk, where she arrived on February 13. She soon began preparations for inactivation. In June, she was moved to Baltimore, Maryland, where she remained until the beginning of 1948. Texas was towed to San Jacinto State Park in Texas, where she was decommissioned on April 21, 1948, and turned over to the state of Texas to serve as a permanent memorial. Her name was struck from the Navy list on April 30, 1948.

The USS Texas (BB-35) earned five battle stars during World War II.

BB-36 • USS NEVADA

(BB-36: dp. 27,500; l. 583'; b. 85'3"; dr. 28'6"; s. 20.5 k.; cpl. 864; a. 10 14", 21 5", 4 21" tt.; cl. Nevada)

The second ship named Nevada (BB-36) was laid down on 4 November 1912 by the Fore River Shipbuilding Co., Quincy, Mass.; launched on 11 July 1914; sponsored by Miss Eleanor Anne Seibert, niece of Governor Tasker L. Oddie of Nevada and descendant of Secretary of the Navy Benjamin Stoddert, and commissioned on 11 March 1916, Capt. William S. Sims in command.

Nevada joined the Atlantic Fleet at Newport on 26 May 1916 and operated along the East Coast and in the Caribbean until World War I. After training gunners out of Norfolk, she sailed on 13 August 1918 to serve with the British Grand Fleet, arriving at Bantry Bay, Ireland, on 23 August. She made a sweep through the North Sea and escorted the transport George Washington, with President Woodrow Wilson embarked, during the last day of her passage into Brest, France, before sailing for home on 14 December.

Nevada served in both the Atlantic and Pacific Fleets in the period between the wars. In September 1922, she represented the United States in Rio de Janeiro for the Centennial of Brazilian Independence. From July to September 1925, she participated in the U.S. Fleet's goodwill cruise to Australia and New Zealand, which demonstrated to our friends down under, and to the Japanese, our ability to make a self-supported cruise to a distance equal to that to Japan. Modernized at Norfolk Naval Shipyard between August 1927 and January 1930, Nevada served in the Pacific Fleet for the next decade.

On 7 December 1941, Nevada was moored singly off Ford Island and had a freedom of maneuver denied to the other 8 battleships present during the attack. As her gunners opened fire and her engineers got up steam, she was struck by one torpedo and two, possibly three, bombs from the Japanese attackers, but was able to get underway. While attempting to leave the harbor, she was struck again. Fearing she might sink in the channel and block it, she was beached at Hospital Point. Gutted forward, she lost 50 killed and 109 wounded.

Refloated on 12 February 1942, Nevada repaired at Pearl Harbor and Puget Sound Navy Yard, then sailed for Alaska where she provided fire support for the capture of Attu from 11 to 18 May. In June, she sailed for further modernization at Norfolk Navy Yard, and in April 1944 reached British waters to prepare for the Normandy Invasion. In action from 6 to 17 June, and again on 25 June, her mighty guns pounded not only permanent shore defenses on the Cherbourg Peninsula but ranged as far as 17 miles inland, breaking up German concentrations and counterattacks. Shore batteries straddled her 27 times, but failed to diminish her accurate fire.

Between 15 August and 25 September, Nevada fired in the invasion of Southern France, dueling at Toulon with shore batteries of 13.4-inch guns taken from French battleships scuttled early in the war. Her gun barrels were relined at New York, and she sailed for the Pacific, arriving off Iwo Jima on 16 February 1945 to give marines invading and fighting ashore her massive gunfire support through 7 March.

On 24 March, Nevada massed off Okinawa with the mightiest naval force ever seen in the Pacific, as pre-invasion bombardment began. She pounded Japanese airfields, shore defenses, supply dumps, and troop concentrations through the crucial operation, although 11 men were killed and a main battery turret damaged when she was struck by a suicide plane on 27 March. Another 2 men were lost to fire from a shore battery on 5 April. Serving off Okinawa until 30 June, from 10 July to 7 August she ranged with the 3rd Fleet which not only bombed the Japanese home islands but came within range for Nevada's guns during the closing days of the war.

Returning to Pearl Harbor after brief occupation duty in Tokyo Bay, Nevada was surveyed and assigned as a target ship for the Bikini atomic experiments. The tough old veteran survived the atom bomb test of July 1946, returned to Pearl Harbor to decommission on 29 August, and was sunk by gunfire and aerial torpedoes off Hawaii on 31 July 1948.

Nevada received 7 battle stars for World War II service.

BB-37 • USS OKLAHOMA

(BB-37: dp. 27,500; l. 583'; b. 95'3"; dr. 28'6"; s. 20.5 k.; cpl. 864; a. 10 14", 20 5", 4 21" tt.; cl. Nevada)

Oklahoma (BB-37) was laid down on October 26, 1912, by the New York Shipbuilding Corp., Camden, New Jersey; launched on March 23, 1914; sponsored by Miss Lorena J. Cruce, and commissioned in Philadelphia on May 2, 1916, with Captain Roger Welles in command.

Joining the Atlantic Fleet with Norfolk as her home port, Oklahoma trained along the eastern seaboard until departing on August 13, 1918, with her sister ship Nevada, to protect Allied convoys in European waters. In December, she was part of the escort for President Woodrow Wilson's arrival in France, leaving on the 14th for New York and winter fleet exercises in Cuban waters. She returned to Brest on June 15, 1919, to escort President Wilson in the George Washington home from his second visit to France, returning to New York on July 8.

A part of the Atlantic Fleet for the next two years, Oklahoma underwent an overhaul, trained, and twice voyaged to South America's west coast; early in 1921 for combined exercises with the Pacific Fleet, and later that year for the Peruvian Centennial. She then joined the Pacific Fleet for six years, highlighted by the Battle Fleet's cruise to Australia and New Zealand in 1925. Joining the Scouting Fleet in early 1927, Oklahoma continued intensive exercises during that summer's Midshipmen Cruise, traveling to the East Coast to embark midshipmen, carrying them through the Panama Canal to San Francisco, and returning via Cuba and Haiti.

Modernized at Philadelphia between September 1927 and July 1929, Oklahoma rejoined the Scouting Fleet for exercises in the Caribbean, returning to the west coast in June 1930 for fleet operations through spring 1936. That summer, she carried midshipmen on a European training cruise, visiting northern ports. The cruise was interrupted by the outbreak of the civil war in Spain, as

Oklahoma rushed to Bilbao, arriving on July 24, 1936, to rescue American citizens and other refugees, whom she carried to Gibraltar and French ports. She returned to Norfolk on September 11, and to the West Coast on October 24.

Oklahoma's Pacific Fleet operations over the next four years included joint operations with the Army and the training of reservists.

Based at Pearl Harbor from December 6, 1940, for patrols and exercises, Oklahoma was moored in Battleship Row on December 7, 1941, when the Japanese attacked. Moored outboard alongside Maryland, Oklahoma took three torpedo hits almost immediately after the first Japanese bombs fell. As she began to capsize, two more torpedoes struck, and her crew was strafed as they abandoned ship. Within 20 minutes of the attack's start, she had capsized, halted by her masts touching the bottom, with her starboard side above water and part of her keel exposed. Many of her crew remained in the fight, boarding Maryland to help serve her anti-aircraft batteries. The attack resulted in twenty officers and 395 enlisted men either killed or missing, 32 others wounded, and many trapped within the capsized hull, saved by heroic rescue efforts. Notably, Julio DeCastro, a civilian yard worker, organized a team that saved 32 Oklahoma sailors.

The salvage operation began in March 1943, and Oklahoma entered dry dock on December 28. Decommissioned on September 1, 1944, Oklahoma was stripped of guns and superstructure and sold on December 5, 1946, to Moore Drydock Co., Oakland, California. Oklahoma parted her tow line and sank on May 17, 1947, 540 miles from Pearl Harbor en route to San Francisco.

Oklahoma received one battle star for World War II service.

BB-38 • USS PENNSYLVANIA

(BB-38: dp. 31,400; l. 608'; b. 97'1"; dr. 28'10"; s. 21 k.; cpl. 915; a. 12 14", 14 5", 4 3", 4 3-pdrs., 2 21" tt.; cl. Pennsylvania)

The second ship named Pennsylvania (BB-38) was laid down on October 27, 1913, by the Newport News Shipbuilding and Dry Dock Co., Newport News, Virginia; launched on March 16, 1915, sponsored by Miss Elizabeth Kolb; and commissioned on June 12, 1916, with Captain H. B. Wilson in command.

Pennsylvania was attached to the Atlantic Fleet. On October 12, 1916, she became the flagship of Commander-in-Chief, U.S. Atlantic Fleet, when Admiral Henry T. Mayo shifted his flag from Wyoming to Pennsylvania. In January 1917, Pennsylvania steamed for fleet maneuvers in the Caribbean and returned to her base at Yorktown, Virginia, on April 6, 1917, the day of the declaration of war against Germany. She did not sail to join the British Grand Fleet since she burned fuel oil, and tankers could not be spared to carry additional fuel to the British Isles.

Based at Yorktown, she maintained battle readiness with fleet maneuvers and tactics in the Chesapeake Bay area, interspersed with overhauls at Norfolk and New York, and brief maneuvers in Long Island Sound.

While at Yorktown on August 11, 1917, Pennsylvania manned the rail and rendered honors as Mayflower, with President Wilson aboard, anchored. At 12:15 p.m., President Wilson returned the call of Commander, Battle Force, aboard Pennsylvania and was given full honors.

On December 2, 1918, Pennsylvania steamed to an anchorage off Tompkinsville, New York. On December 4, she got underway for Brest, France. At 11:00 a.m., transport George Washington, flying the flag of the President of the United States, set out with an escort of ten destroyers. Pennsylvania manned the rail and fired a 21-gun salute. She took position ahead of George Washington as the guide for the President's escort. Arriving in Brest on December 13, the crew manned the rail and cheered as George Washington passed and proceeded to her anchorage. On December 14, Pennsylvania departed for New York, arriving on December 25.

In February 1919, Pennsylvania steamed for fleet maneuvers in the Caribbean Sea and returned to New York in late spring. While in New York on June 30, 1919, Admiral Mayo was relieved as Commander-in-Chief, U.S. Atlantic Fleet, by Vice Admiral Henry B. Wilson.

At Tompkinsville, New York, on July 8, 1919, Pennsylvania embarked Vice President Marshall, Cabinet Secretaries Daniels, Glass, Wilson, Baker, Lane, and Senator Champ Clark, and then put to sea. At 10:00 a.m., Oklahoma was sighted with George Washington, flying the President's flag, and accompanied by her ocean escort. Pennsylvania fired a presidential salute, then took position ahead of Oklahoma and steamed to New York, stopping en route to disembark her distinguished guests before proceeding to her berth.

On January 7, 1920, she departed New York for fleet maneuvers in the Caribbean Sea, returning to New York on April 26, 1920. She resumed a schedule of local training operations until January 17, 1921, when she departed New York for the Panama Canal, arriving at Balboa on January 20. There, she joined units of the Pacific Fleet and became the flagship of the combined fleets, with the Commander-in-Chief, U.S. Atlantic Fleet, assuming command of the U.S. Battle Fleet. On January 21, 1921, the fleet sailed from Balboa to Callao, Peru, arriving on January 31. Departing on February 2, Pennsylvania returned to Balboa on February 14, then conducted brief exercises while based at Guantanamo Bay, Cuba. Upon returning to Hampton Roads on April 28, 1921, she rendered a 21-gun salute as she passed Mayflower. The Secretary of the Navy, the Chief of Naval Operations, and the Assistant Secretary of the Navy came aboard for a reception for President Harding. At 11:40 a.m., President Harding came aboard and his flag was broken at the main.

On August 22, 1922, Pennsylvania departed Hampton Roads to join the Pacific Fleet. Arriving in San Pedro, California, on September 26, 1922, her principal area of operations until 1929 was along the coast of California, Washington, and Oregon, with periodic maneuvers and tactics off the Panama Canal, in the Caribbean Sea, and Hawaiian operating areas. She departed with the fleet from San Francisco on April 15, 1925, and, after war games in the Hawaiian area, departed Honolulu on July 1 en route to Melbourne, Australia. After visiting Wellington, New Zealand, she returned to San Pedro on September 26, 1925.

In January 1929, Pennsylvania cruised to Panama, and after training maneuvers based at Guantanamo Bay, Cuba, steamed to the Philadelphia Navy Yard, arriving on June 1, 1929, for an overhaul and modernization. She remained in the yard for nearly two years. On May 8, 1931, she departed for a refresher training cruise to Guantanamo Bay, Cuba, and then returned. On August 6, 1931, she again sailed for Guantanamo and later continued to San Pedro, where she rejoined the Battle Fleet.

From August 1931 to 1941, Pennsylvania engaged in fleet tactics and battle practice along the west coast and participated in fleet problems and maneuvers held periodically in the Hawaiian area and the Caribbean Sea. After an overhaul in the Puget Sound Naval Shipyard on January 7, 1941, she sailed for Hawaii, where she carried out scheduled operations with units of Task Forces 1 and 5 throughout the year, making one brief voyage to the west coast with Task Force 18.

At the time of the Japanese attack on Pearl Harbor on December 7, 1941, Pennsylvania was in dry dock in the Pearl Harbor Navy Yard. She was one of the first ships in the harbor to open fire as enemy dive bombers and torpedo planes appeared. Despite repeated attempts, they did not succeed in torpedoing the caisson of the dry dock, but Pennsylvania and the surrounding dock areas were severely strafed. The crew of one 5-inch gun mount was wiped out when a bomb struck the starboard side of her boat deck and exploded inside casemate 9. Destroyers Cassin and Downes, just forward of Pennsylvania in dry dock, were seriously damaged by bomb hits. Pennsylvania was pockmarked by flying fragments. A part of a torpedo tube from destroyer Downes, weighing about 1000 pounds, was blown onto the forecastle of Pennsylvania. She suffered 15 men killed, 14 missing in action, and 38 men wounded.

On December 20, 1941, Pennsylvania sailed for San Francisco, arriving on December 29. She underwent repairs until March 30, 1942. From April 14 to August 1, 1942, Pennsylvania conducted extensive training operations and patrols along the coast of California, interspersed with an overhaul in San Francisco. During this duty, on June 4, 1942, Admiral Ernest J. King, Commander-in-Chief of the United States Fleet, held brief ceremonies aboard Pennsylvania to present the Distinguished Service Medal to Admiral Chester W. Nimitz for exceptionally meritorious service as Commander-in-Chief of the U.S. Pacific Fleet since December 31, 1941.

On August 1, 1942, Pennsylvania departed San Francisco for Pearl Harbor, arriving on August 14. She conducted gunnery exercises and took part in carrier task force guard tactics in the Hawaiian area. On October 4, Pennsylvania returned to San Francisco for an overhaul completed by February 5, 1943. She then

conducted refresher training and air defense patrol off the coast of California. On April 23, Pennsylvania sailed for Alaska to take part in the Aleutian Campaign.

On April 30, Pennsylvania arrived at Cold Bay, Alaska. On May 11-12, she engaged in shore bombardment of Holtz Bay and Chicago Harbor, Attu, in support of the landings. As she retired from Attu on May 12, a patrol plane warned of a torpedo heading towards Pennsylvania. She maneuvered at full speed as the torpedo passed safely astern. Destroyers Edwards and Farragut hunted down the attacker. After ten hours of relentless depth charge attacks, submarine I-81 was forced to the surface and shelled by gunfire from Edwards. Severely damaged, the enemy survived until June 13, then being sunk by destroyer Frazier. Torpedo wakes were again sighted on the morning of May 14, but the destroyers' search for the enemy was fruitless. That same morning, Pennsylvania's seaplanes were launched to operate from seaplane tender Casco in making strafing attacks on enemy positions on Attu.

The afternoon of 14 May, Pennsylvania conducted her third bombardment mission, this time in support of the infantry attack on the west arm of Holtz Bay. She then operated to the north and east of Attu until 19 May when she steamed for Adak. She departed Adak 21 May and arrived at the Puget Sound Navy Yard, Bremerton, Wash., 28 May. She returned to Adak, 7 August, and departed 13 August as flagship of Admiral Rockwell, commanding the Kiska Attack Force. On 15 August assault troops landed without opposition on the western beaches of Kiska. By the evening

of 16 August it became apparent the Japanese had evacuated under cover of fog prior to the landing. She patrolled off Kiska for a time then returned to Adak, 23 August.

On 25 August Pennsylvania steamed for Pearl Harbor, arriving 1 September. Here she took aboard 790 passengers and departed 19 September for San Francisco where she arrived 25 September. She returned to Pearl Harbor, 6 October, and after debarking passengers, took part in rehearsal and bombardment exercises in the Hawaiian areas. She became flagship of Rear Admiral Richmond K. Turner, Commander Fifth Amphibious Force, and formed part of the Northern Attack Force, departing Pearl Harbor, 10 November, for the assault on Makin Atoll, Gilbert Islands.

The Task Force, comprising four battleships, four cruisers, three escort carriers, transports and destroyers, approached Makin Atoll from the southeast on the morning of 20 November. Pennsylvania opened fire on Butaritari Island with her main battery at the initial range of 14,200 yards and then opened with her secondary battery.

Just before general quarters on the morning of 24 November a tremendous explosion took place off the starboard bow as Pennsylvania was returning to a screening sector off Makin. At almost the same instant a screening destroyer reported sound contact and disposition immediately executed a course change. For several minutes after the explosion, a large fire lighted up the entire area. Word soon came that escort carrier Liscome Bay had been torpedoed. She sank with tremendous loss of life. Determined night air attacks were made by enemy torpedo planes on the nights of 25 and 26 November but were repelled without damage to ships of the Task Force.

On 31 January 1944, Pennsylvania commenced bombardment of Kawjalein Island which was continued throughout the day. Landings were made 1 February, with Pennsylvania joining in bombardment support before and after the landing operations. On the evening of 3 February, she anchored in the lagoon near Kwajalein Island. The success of the Kwajalein operation was ensured and Pennsylvania retired to Majuro Atoll to replenish ammunition.

On 12 February Pennsylvania got underway for operations against Eniwetok, Marshall Islands. On 17

February, Pennsylvania steamed boldly through the deep entrance into Eniwetok Lagoon with her batteries blazing away. She steamed up a swept channel in the lagoon to a position off Engebi Island and commenced bombardment of enemy installations. On the morning of 18 February, Pennsylvania bombarded Engebi before and during the approach of the assault waves to the beach. When Engebi had been secured, Pennsylvania steamed southward through the lagoon to the vicinity of Parry Island, where she took part in bombardment 20-21 February, preparatory to the landing assaults. At the commencement of bombardment the island had been covered with a dense growth of palm trees extending to the waters edge. At conclusion of bombardment, not a single tree remained standing. On the morning of 22 February, she gave bombardment support prior to the landing on Parry Island.

Pennsylvania retired to Majuro, 1 March, then steamed south to Havannah Harbor, Efate, New Hebrides Islands. She remained at Efate until late April. On 29 April, Pennsylvania arrived in Sydney, Australia. She returned to Efate, 11 May, then sailed to Port Purvis, Florida Islands, from which she operated to conduct bombardment and amphibious assault exercises. She returned to Efate 27 March, and after replenishment of ammunition, departed, 2 June, arriving at Roi, 3 June.

On 10 June, Pennsylvania formed with a force of battleships, cruisers, escort carriers, and destroyers en route for the assault and occupation of the Marianas Islands. That night a destroyer in the screen reported sound contact and emergency turn left 90 degrees was ordered. As a result of this maneuver, Pennsylvania collided with high speed transport Talbot and sustained minor damage. Talbot put into Eniwetok for emergency repairs.

On 14 June, Pennsylvania took part in the bombardment of Saipan preparatory to the assault landings made the next

day while she cruised off the northeastern shore of Tinian conducting heavy bombardment of that island to neutralize any enemy batteries which might have opened fire on the landing beaches of Saipan. On 16 June she conducted bombardment of targets on Orote Point, Guam, then retired to cover the Saipan area. Pennsylvania departed the Marianas, 25 June, and after a brief stay at Eniwetok, Marshall Islands, departed 9 July to resume support of the Marianas Campaign.

From 12 through 14 July, Pennsylvania conducted bombardment of Guam in preparation for the assault and landings on that island. On completion of firing the evening of 14 July she returned to Saipan to replenish ammunition. She returned to Guam, 17 July, and delivered protective fire support to demolition parties. At the same time she continued deliberate destructive fire on designated targets through 20 July.

On the early morning of 21 July, Pennsylvania took a position between Agat Beach and Orote Peninsula, and commenced bombardment of beach areas in immediate preparation for the assault while troops and equipment were loaded into landing craft and landing waves were being formed. Upon establishment of the beachhead she stood by for fire support missions as might be called for by shore fire control parties, continuing this duty until 3 August. She then steamed to Eniwetok, thence to the New Hebrides Islands, and after rehearsal of landing assaults on Cape Esperance, Guadalcanal, arrived at Port Purvis, Florida Island. She departed 6 September as part of the Palau Bombardment and Fire Support Group. From 12 through 14 September, Pennsylvania took part in intensive bombardment of targets on the island of Peleliu. On 15 September, she also furnished gunfire support for the landings on that island. She then delivered a devastating fire on enemy gun emplacements among the rocks and cliffs flanking Red Beach on Angaur Island.

On 25 September Pennsylvania steamed for emergency repairs at Manus Admiralty Island, entering floating drydock, 1 October 1944. She departed 12 October, one of six battleships in Rear Admiral Jesse B. Oldendorf's Bombardment and Fire Support Group which formed a part of the Central Philippine Attack Force under command of Vice Admiral Thomas Cassin Kinkaid, en route to the Philippine Islands.

Pennsylvania reached fire support station on the eastern coast of Leyte, 18 October, and commenced covering bombardment for beach reconnaissance, underwater demolition teams and minesweeping units operating in Leyte Gulf and San Pedro Harbor. She conducted bombardment missions the next day and supported the landings on Leyte, 20 October. Gunfire support missions continued through 22 October, including harrassing and night illumination fire.

On 24 October, all available United States vessels prepared for action as units of the Japanese Fleet closed the Philippines preliminary to the Battle for Leyte Gulf. Pennsylvania and five other battleships, with cruisers and destroyers of Rear Admiral Oldendorf's Force, steamed south and by nightfall were steaming slowly back and forth across the northern entrance of Surigao Strait, awaiting the approach of the enemy. That night, American motor torpedo boats stationed well down in Surigao Strait made the first encounter with torpedo attacks. Destroyers of the Force, on either flank of the enemy's line of approach, followed with torpedo and gun attacks. At 0353, 25 October, West Virginia opened fire, joined shortly thereafter by other battleships and cruisers. The Japanese had run head on into a perfect trap. Rear Admiral Oldendorf had executed the dream of every naval taetieian by crossing the enemy's "T". The Japanese lost two battleships and three destroyers in the Battle of Surigao Strait. Cruiser Mogami in company with a destroyer, all that remained of the enemy force, managed to escape. Rear Admiral Oldendorf's Force did not suffer the loss of a single vessel. Moyami was sunk the next day by carrier planes.

On 25 October 1944 ten enemy planes made a simultaneous run on a destroyer close aboard Pennsylvania which assisted in splashing four of the others. On the night of 28 October, she shot down a bomber as it attempted a torpedo run.

Remaining on patrol in Leyte Gulf until November 25, the Pennsylvania then steamed to Manus, Admiralty Islands, and subsequently to Kossol Passage to load ammunition. On January 1, 1945, it departed with Vice Admiral Oldendorf's Lingayen Bombardment and Fire Support Group, heading for Lingayen Gulf. The Group encountered heavy air attacks on January 4-5, and the escort carrier Ommaney Bay was hit by a kamikaze plane and destroyed by the resulting fire. Many other ships were damaged.

On the morning of January 6, the Pennsylvania began bombarding target areas on Santiago Island at the mouth of Lingayen Gulf. That afternoon, it entered the Gulf to conduct counter-battery fire in support of minesweeping forces, retreating at night. At daybreak on January 7, the entire bombardment force entered Lingayen Gulf to deliver supporting and destructive fire. Preliminary assault bombardment continued the next day. On January 9, the Pennsylvania provided gunfire support to protect the landing troops. Enemy aircraft attacked the force in Lingayen Gulf on January 10. Four bombs landed nearby, but the Pennsylvania was not hit. That afternoon, she executed her last call fire mission of the operation, firing twelve rounds to destroy a concentration of enemy tanks located inland by a shore fire control party.

From January 10 to 17, the Pennsylvania patrolled the South China Sea off Lingayen Gulf with other ships of the task group. On January 17, she anchored in Lingayen Gulf, remaining there until February 10, when she sailed for temporary repairs at Manus, Admiralty Islands. Departing on February 22, she steamed via the Marshall Islands and Pearl Harbor to San Francisco, arriving on March 13. She underwent a thorough overhaul at the Hunter's Point Shipyard. Her main battery turrets and secondary battery mounts were regunned. Additional close-range weapons, improved radar, and fire control equipment were installed.

Upon completion of the overhaul, the Pennsylvania conducted trial runs out of San Francisco, followed by refresher training while based in San Diego, California. She departed San Francisco on July 12 for Pearl Harbor, arriving on July 18. She sailed for Okinawa on July 24. En route, she participated in the bombardment of Wake Island on August 1, and after loading ammunition at Saipan the next day, resumed her voyage. She anchored in Buckner Bay alongside Tennessee. On August 12, a Japanese torpedo plane evaded detection and launched a torpedo at the anchored Pennsylvania, causing extensive damage. The attack killed twenty men and injured ten. Many compartments were flooded, causing the ship to settle heavily by the stern. The flooding was controlled by the Pennsylvania's repair parties and the prompt assistance of two salvage tugs. The next day, she was towed to shallower waters for continued salvage operations.

On August 18, the Pennsylvania departed Buckner Bay, Okinawa, under the tow of two tugs. She arrived at Apra Harbor, Guam, on September 6, and entered drydock where a large steel patch was welded over the torpedo hole, and repairs were made to enable her return to the United States under her own power. On October 4, she sailed for the Puget Sound Navy Yard, accompanied by the destroyer Walke and cruiser Atlanta. On October 17, the number 3 shaft broke inside the stern tube and slipped aft. Divers had to cut through the shaft, allowing it and the propeller to drop into the sea. With only one functioning screw and taking on water, the Pennsylvania limped into the Puget Sound Navy Yard on October 24.

Repairs enabled the Pennsylvania to steam to the Marshall Islands, where she was used as a target ship in the atomic bomb tests at Bikini Atoll in July 1946. Afterwards, she was towed to Kwajalein Lagoon, where she was decommissioned on August 29, 1946. She remained in Kwajalein Lagoon for radiological and structural studies until February 10, 1948, when she was sunk off Kwajalein. She was struck from the Navy List on February 19, 1948.

The Pennsylvania received eight battle stars for her service in World War II.

BB-39 • USS ARIZONA

(Battleship No. 39: dp. 31,400; 1. 608"; b. . 97'1"; dr. 28'10" (mean); s. 21 k.; cpl. 1,081; a. 12 14", 22 5", 4 3", 2 21" tt.; cl. Pennsylvania)

The second ship named Arizona (Battleship No. 39) was laid down on 16 March 1914 at the New York Navy Yard; launched on 19 June 1915, sponsored by Miss Esther Ross, daughter of a prominent Arizona pioneer citizen, Mr. W. W. Ross of Prescott, Ariz.; and commissioned at her builder's yard on 17 October 1916, Capt. John D. McDonald in command.

Arizona departed New York on 16 November 1918 for shakedown training off the Virginia Capes and Newport, proceeding thence to Guantanamo Bay, Cuba. Returned north to Norfolk on 16 December to test fire her battery and to conduct torpedo-defense exercises in Tangier Sound. The battleship returned to her builder's yard the day before Christmas of 1916 for post-shakedown overhaul. Completing these repairs and alterations on 3 April 1917,

she cleared the yard for Norfolk on that date, arriving there on the following day to join Battleship Division 8.

Within days, the United States forsook its tenuous neutrality in the global conflict, then raging and entered World War I. The new battleship operated out of Norfolk throughout the war, serving as a gunnery training ship and patrolling the waters of the eastern seaboard from the Virginia Capes to New York. An oil burner, she had not been deployed to European waters due to a scarcity of fuel oil in the British Isles, the base of other American battleships sent to aid the Grand Fleet.

A week after the armistice of 11 November 1918 stilled the guns on the western front, Arizona stood out of Hampton Roads for Portland, England, and reached her destination on 30 November 1918, putting to sea with her division on 12 December to rendezvous with the transport George Washington, the ship carrying President Woodrow Wilson to the Paris Peace Conference. Arizona, one of the newest and most powerful American dreadnoughts, served as part of the honor escort convoying the American President to Brest, France, on 13 December 1918.

Embarking 238 homeward-bound veterans in the precursor of a "Magic Carpet" operation of a later war, Arizona sailed from Brest for New York on 14 December and arrived off Ambrose Light on the afternoon of Christmas Day, 1918. The next day, she passed in review before Secretary of the Navy Josephus Daniels, who was embarked in the yacht Mayflower off the Statue of Liberty, before entering New York Harbor in a great homecoming celebration. The battleship then sailed for Hampton Roads on 22 January 1919, returning to her base at Norfolk on the following day. Arizona sailed for Guantanamo Bay with the Fleet on 4 February 1919 and arrived on the 8th. After engaging in battle practices and maneuvers there, the battleship sailed for Trinidad on 17 March, arriving there five days later or a three-day port visit. She then returned to Guantanamo Bay on 29 March or a brief period, sailing for Hampton Roads on 9 April. Arriving at her destination on the morning of the 12th, she got underway late that afternoon for Brest, France, ultimately making arrival there on 21 April 1919.

The battleship stood out of Brest harbor on 3 May, bound for Asia Minor, and arrived at the port of Smyrna eight days later to protect American lives there during the Greek occupation of that port occupation, resisted by gunfire from Turkish nation ls. Arizona provided temporary shelter on board for a party of Greek nationals while the battleship's marine detachment guarded the American consulate; a number of American citizens also remained on board Arizona until conditions permitted them to return ashore. Departing Smyrna on 9 June for Constantinople, Turkey, the battleship carried the United States consul-at-large, Leland E. Morris, to that port before sailing for New York on 15 June. Proceeding via Gibraltar, Arizona, she reached her destination on 30 June.

Entering the New York Navy Yard for upkeep soon thereafter, the battleship cleared that port on 6 January 1920 to join Battleship Division 7 for winter and spring maneuvers in the Caribbean. She operated out of Guantanamo Bay during this period and also visited Bridgetown, Barbados, in the British West Indies, and Colon, in the Canal Zone, before she sailed north for New York, arriving there on 1

May 1920. Departing New York on May 17, Arizona operated on the Southern Drill Grounds and then visited Norfolk and Annapolis before returning to New York on June 25. Over the next six months, the ship operated locally out of New York. During this time, she was given the alphanumeric hull designation BB-39 on 17 July 1920, and on 23 August, she became a flagship for Commander Battleship Division 7, Rear Admiral Edward V. Eberle. Sailing from New York on 4 January 1921, Arizona joined the feet as it sailed for Guantanamo Bay and the Panama Canal Zone. Arriving at Colon, on the Atlantic side of the Isthmian waterway, on 19 January, Arizona transited the Panama Canal for the first time, arriving at Panama Bay on the 20th. Underway for Callao, Peru, on the 22nd, the fleet arrived there nine days later, on the three 1st, for a six-day visit. While she was there, Arizona was visited by the President of Peru. Underway for Balboa on 5 February 1921, Arizona arrived at the destination on the 14th; transiting the canal again the day after Washington's birthday, the battleship reached Guantanamo Bay on the 6th. She operated thence until 24 April 1921, when she sailed for New York, steaming via Hampton Roads.

Arizona reached New York on 29 April and remained under overhaul there until 15 June. She steamed thence for Hampton Roads on the latter date and, on the 21st, steamed off Cape Charles with Army and Navy observers to witness the experimental bombings of the ex-German submarine U-117. Proceeding thence back to New York, the battleship there broke the flag of Vice Admiral John D. McDonald (who, as a captain, had been Arizona's first commanding officer) on 1 July and sailed for Panama and Peru on 9 July. She arrived at the port of Callao on 22 July as flagship for the Battle Force, Atlantic Fleet, to observe the celebrations accompanying the centennial year of Peruvian independence. On 27 July, Vice Admiral McDonald went ashore and represented the United States at the unveiling of a monument commemorating the accomplishments of San Martin, who had liberated Peru from the Spanish yoke a century before.

Sailing for Panama Bay on 9 August, Arizona became the flagship for Battleship Division 7 when Vice Admiral McDonald transferred his flag to Wyoming (BB-33), and Rear Admiral Josiah S. McKean broke his flag on board as commander of the division on 10 August at Balboa. The following day, the battleship sailed for San Diego, arriving there on 21 August 1921.

Over the next 14 years, Arizona alternately served as flagship for Battleship Divisions 2, 9, or 4. Based at San Pedro, during this period, Arizona operated with the fleet in the operating areas off the coast of southern California or in the Caribbean during fleet concentrations there. She participated in a succession of fleet problems (the annual maneuvers of the fleet that served as the culmination of the training year), ranging from the Caribbean to the waters off the west coast of Central America and the Canal Zone, from the West Indies to the waters between Hawaii and the west coast.

Following her participation in Fleet Problem IX (January 1929), Arizona transited the Panama Canal on 7 February for Guantanamo Bay, whence she operated

through April. She then proceeded to Norfolk Navy Yard, entering it on 4 May 1929 to prepare for modernization.

Placed in reduced commission on 15 July 1929, Arizona remained in yard hands for the next 20 months; tripod masts, surmounted by three-tiered fire control tops, replaced the old cage masts; 5-inch, 25-caliber antiaircraft guns replaced the 3-inch, 50s with which she had been equipped. She also received additional armor to protect her vitals from the fall of shot and blisters to protect her from torpedoes or near-miss damage from bombs. In addition, she received new boilers as well as new main and cruising turbines. Ultimately, she was placed in full commission on 1 March 1931.

A little over two weeks later, on 19 March 1931, President Herbert C. Hoover embarked on board the recently modernized battleship and sailed for Puerto Rico and the Virgin Islands, standing out to sea from Hampton Roads that day. Returning on 29 March, Arizona disembarked the Chief Executive and his party at Hampton Roads and then proceeded north to Rockland, Maine, to run her post-modernization standardization trials. After a visit to Boston, the battleship dropped down to Norfolk, where she sailed for San Pedro on 1 August 1931, assigned to Battleship Division 3, Battle Force.

Over the next decade, Arizona continued to operate with the Battle Fleet and took part in the succession of fleet problems that took the fleet from the waters of the northern Pacific and Alaska to those surrounding the West Indies and into the waters east of the lesser Antilles.

On 17 September 1938, Arizona became the flagship for Battleship Division 1 when Rear Admiral Chester W. Nimitz (later to become Commander-in-Chief, Pacific Fleet) broke his flag on board. Detached 27 May 1939 to become Chief of the Bureau of Navigation, Nimitz was relieved on that day by Rear Admiral Russell Willson.

Arizona's last fleet problem was XXI. At its conclusion, the United States Fleet was retained in Hawaiian waters, based at Pearl Harbor. She operated in the Hawaiian Operating Area until late that summer when she returned to Long Beach on 30 September 1940. She was then overhauled at the Puget Sound Navy Yard, Bremerton, Wash., into the following year. Her last flag change of command occurred on 23 January 1941, when Rear Admiral Wilson was relieved as Commander, Battleship Division 1 by Rear Admiral Isaac C. Kidd.

The battleship returned to Pearl Harbor on February 3, 1941, to resume the intensive training maintained by the Pacific Fleet. She made one last visit to the West Coast, clearing "Pearl" on 11 June 1941 for Long Beach, ultimately returning to her Hawaiian base on 8 July. Over the next five months, she continued exercises and battle problems of various kinds of training and tactical exercises in the Hawaiian operating area. She underwent a brief overhaul at the Pearl Harbor Navy Yard commencing on 27 October 1941, receiving the foundation for a search radar atop her foremast. She conducted her last training in company with her division mates Nevada (BB-36) and Oklahoma (BB-37), conducting a night firing exercise on the night of 4 December 1941. All three ships moored at quays ("keys") along Ford Island on the 5th.

Scheduled to receive tender availability, Arizona took the repair ship Vestal (AR-4) alongside on Saturday, the 6th. The two ships were thus moored together on the morning of 7 December; among the men on board Arizona, that morning were Rear Admiral Kidd and the battleship's captain, Capt. Franklin van Valkenburgh.

Shortly before 0800, Japanese aircraft from six fleet carriers struck the Pacific Fleet as it lay in port at Pearl Harbor, and in the ensuing two attack waves, wrought devastation on the Battle Line and on air and military facilities defending Pearl Harbor.

On board Arizona, the ship's air raid alarm went off at about 0755, and the ship went to general quarters soon thereafter. Insofar as it could be determined soon after the attack, the ship sustained eight bomb hits; one hit on the forecastle, glancing off the face plate of turret II to penetrating the deck to explode in the black powder magazine, which in turn set off adjacent smokeless powder magazines. A cataclysmic explosion ripped through the forward part of the ship, touching off fierce fires that burned for two days; debris showered down on Ford Island in the vicinity.

Acts of heroism on the part of Arizona's officers and men were many, headed by those of Lt. Comdr. Samuel

G. Fuqua, the ship's damage control officer, whose coolness in attempting to quell the fires and get survivors off the ship earned him the Medal of Honor. Posthumous awards of the Medal of Honor also went to Rear Admiral Isaac Kidd, the first flag officer to be killed in the Pacific war, and to Capt. Van Valkenburgh reached the bridge and was attempting to fight his ship when the bomb hit the magazines, destroying her.

The blast that destroyed Arizona and sank her at her berth alongside Ford Island consumed the lives of 1,103 of the 1,400 on board, at the time-over, half of the casualties suffered by the entire fleet on the "Day of Infamy."

Placed "in ordinary" at Pearl Harbor on 29 December 1941, Arizona was struck from the Naval Vessel Register on 1 December 1942. Her wreck was cut down so that very little of the superstructure lay above water; after the main battery turrets, guns were removed to be emplaced as coast defense guns. Arizona's wreck remains at Pearl Harbor, a memorial to the men of her crew lost that December morning in 1941. On 7 March 1950, Admiral Arthur W. Radford, Commander in Chief of the Pacific Fleet at that time, instituted the raising of colors over Arizona's remains, and legislation during the administrations of Presidents Dwight D. Eisenhower and John F. Kennedy designated the wreck a national shrine. A memorial was built; it was dedicated on 30 May 1962.

Arizona (BB-39) was awarded one battle star for her service in World War 11.

BB-40 • USS NEW MEXICO

(BB-40: dp. 32,000; l. 624'; b. 97'; dr. 30'; s. 21 k.; cpl. 1,084; a. 12 14", 14 5", 4 3", 2 21" tt.; cl. New Mexico)

The New Mexico (BB-40) was laid down on October 14, 1915, by the New York Navy Yard. Launched on April 13, 1917, she was sponsored by Miss Margaret C. DeBaca, daughter of the Governor of New Mexico, and commissioned on May 20, 1918, with Captain Ashley H. Robertson in command.

After initial training, New Mexico departed New York on January 15, 1919, for Brest, France, to escort the transport George Washington, carrying President Woodrow Wilson from the Versailles Peace Conference, returning to Hampton Roads on February 27. There, on July 16, she became the flagship of the newly-organized Pacific Fleet and three days later sailed for the Panama Canal and San Pedro, California, arriving on August 9. The next 12 years were marked by frequent combined maneuvers with the Atlantic Fleet in the Pacific and Caribbean, including visits to South American ports and a 1925 cruise to Australia and New Zealand.

Modernized and overhauled at Philadelphia between March 1931 and January 1933, New Mexico returned to the Pacific in October 1934 to resume training exercises and tactical development operations. As war loomed, she was based in Pearl Harbor from December 6, 1940, until May 20, 1941, when she sailed to join the Atlantic Fleet at Norfolk on June 16 for duty on neutrality patrol. After the Japanese attack on Pearl Harbor, she returned to the West Coast and sailed from San Francisco on August 1, 1942, to prepare in Hawaii for action. Between December 6, 1942, and March 22, 1943, she escorted troop transports to the Fijis, patrolled the Southwest Pacific, and returned to Pearl Harbor to prepare for the Aleutian campaign. Arriving in Adak on May 17, she served on the blockade of Attu and joined the massive bombardment of Kiska on July 21, forcing its evacuation a week later.

After refitting at Puget Sound Navy Yard, New Mexico returned to Pearl Harbor on October 25 to rehearse for the Gilbert Islands assault. During the invasion, which began on November 20, she bombarded Butaritari, guarded transports during their night withdrawals, and provided anti-aircraft cover during unloading operations, as well as screening carriers. She returned to Pearl Harbor on December 5.

Underway with the Marshall Islands assault force on January 12, 1944, New Mexico bombarded Kwajalein and Ebeye on January 31 and February 1, then replenished at Majuro. She blasted Wotje on February 20 and Kavieng, New Ireland, on March 20, then visited Sydney before arriving in the Solomons in May to rehearse for the Marianas operation.

New Mexico bombarded Tinian on June 14, Saipan on June 15, and Guam on June 16, and twice helped repel enemy air attacks on June 18. She protected transports off the Marianas while the carrier task force secured a major victory, the Battle of the Philippine Sea, on June 19-20. New Mexico escorted transports to Eniwetok, then sailed on July 9, guarding escort carriers until July 12, when her guns opened on Guam in preparation for the landings on July 21. She bombarded enemy positions and installations on the island until July 30.

After an overhaul at Bremerton from August to October, New Mexico arrived in Leyte Gulf on November 22 to cover the movement of reinforcement and supply convoys, firing in the almost daily air attacks over the Gulf, as the Japanese fiercely resisted the reconquest of the Philippines. She left Leyte Gulf on December 2 for the Palau Islands, where she joined a force covering the Mindoro-bound assault convoy. Again, she provided anti-aircraft cover as invasion troops stormed ashore on December 15, remaining for two days before sailing for the Palaus.

Her next operation was the invasion of Luzon, fought under skies filled with would-be suicide planes. She fired pre-landing bombardments on January 6, 1945, and sustained a suicide hit on her bridge that day, killing her commanding officer, Captain R. W. Fleming, and 29 others of her crew, with 87 injured. Her guns remained in action as she repaired damage, and she continued to support the landings on January 9.

After repairs at Pearl Harbor, New Mexico arrived at Ulithi to stage for the invasion of Okinawa, sailing on March 21 with a heavy fire support group. Her guns opened on Okinawa on March 26, and she continued supporting troops ashore until April 17. She resumed fire on April 21 and 29 and on May 11 destroyed eight suicide boats. On May 12, while approaching her berth in Hagushi anchorage, New Mexico was attacked by two kamikazes; one crashed into her, and the other hit her with a bomb. The attack killed 54 of her men and wounded 119. Swift action extinguished the fires within half an hour, and on May 28, she departed for repairs at

Leyte, followed by rehearsals for the planned invasion of the Japanese home islands. She learned of the war's end at Saipan on August 15 and sailed for Okinawa the next day to join the occupation force. She entered Sagami Wan on August 27 to support the airborne occupation of Atsugi Airfield, then moved into Tokyo Bay on August 28 to witness the surrender on September 2.

New Mexico was homeward bound on September 6, calling at Okinawa, Pearl Harbor, and the Panama Canal before arriving in Boston on October 17. There, she was decommissioned on July 19, 1946, and sold for scrapping to Lipsett, Inc., New York City, on October 13, 1947.

New Mexico received six battle stars for her World War II service.

BB-41 • USS MISSISSIPPI

(BB-41: dp. 32,000; l. 624'; b. 97'6"; dr. 30'; s. 21 k.; cpl. 1,081; a. 12 14", 14 5", 4 3", 2 21" tt.; cl. New Mexico)

Mississippi (BB-41) was laid down on 5 April 1915 by the Newport News Shipbuilding Co., Newport News, VA; launched on 25 January 1917; sponsored by Miss Camelle McBeath; and commissioned on 18 December 1917 with Capt. J. L. Jayne in command.

Following exercises off Virginia, Mississippi steamed on 22 March 1918 for training in the Gulf of Guacanayabo, Cuba. One month later, she returned to Hampton Roads and cruised between Boston and New York until departing for winter maneuvers in the Caribbean on 31 January 1919. On 19 July, she left the Atlantic seaboard and sailed for the west coast. Arriving at her new base, San Pedro, she operated along the west coast for the next four years, entering the Caribbean during the winter months for training exercises.

During gunnery practice on 12 June 1924 off San Pedro, 48 of her men were asphyxiated as a result of an explosion in her No. 2 main battery turret. On 15 April 1925, she sailed from San Francisco for war games off Hawaii and then steamed to Australia on a goodwill tour. She returned to the west coast on 26 September and resumed operations there for the next ten years. During this period, she frequently sailed into Caribbean and Atlantic waters for exercises during the winter months.

Mississippi entered the Norfolk Navy Yard on 30 March 1931 for a modernization overhaul, departing once again for training exercises in September 1933. Transiting the Panama Canal on 24 October 1934, she steamed back to her base at San Pedro. For the next seven years, she operated off the west coast, except for winter Caribbean cruises.

Returning to Norfolk on 15 June 1941, she prepared for patrol service in the North Atlantic. Steaming from Newport, RI, she escorted a convoy to Hvalfjordur, Iceland.

She made another trip to Iceland on 28 September 1941 and spent the next two months there protecting shipping.

Two days after the treacherous attack on Pearl Harbor, Mississippi left Iceland for the Pacific.

Arriving on 22 January 1942 at San Francisco, she spent the next seven months training and escorting convoys along the coast. On 6 December, after participating in exercises off Hawaii, she steamed with troop transports to the Fiji Islands, returning to Pearl Harbor on 2 March 1943. On 10 May, she sailed from Pearl Harbor to participate in a move to restore the Aleutians to their rightful possessors. Kiska Island was shelled on 22 July, and a few days later the Japanese withdrew. After an overhaul at San Francisco, Mississippi sailed from San Pedro on 19 October to take part in the invasion of the Gilbert Islands. While bombarding Makin on 20 November, a turret explosion, almost identical to the earlier tragedy, killed 43 men.

On 31 January 1944, she took part in the Marshall Islands campaign, shelling Kwajalein. She bombarded Taroa on 20 February and struck Wotje the next day. On 15 March, she pounded Rabaul, New Ireland. Due for an overhaul, she spent the summer months at Puget Sound.

Returning to the war zone, Mississippi supported landings on Peleliu in the Palau Islands on 12 September. After a week of continuous operations, she steamed to Manus, where she remained until 12 October. Departing Manus, she assisted in the liberation of the Philippines, shelling the east coast of Leyte on 19 October. On the night of the 24th, as part of Admiral Oldendorf's battle line, she helped to destroy a powerful Japanese task force at the Battle of Surigao Strait. As a result of the engagements at Leyte Gulf, the Japanese navy was no longer able to mount any serious offensive threat.

Mississippi continued to support the operations at Leyte Gulf until 16 November, when she steamed to the Admiralty Islands. She then entered San Pedro Bay, Leyte, on 28 December to prepare for the landings on Luzon. On 6 January 1945, she began bombarding Lingayen Gulf. Despite damages near her waterline received from the crash of a suicide plane, she supported the invasion forces until 10 February. Following repairs at Pearl Harbor, she sailed to Nakagusuku Wan, Okinawa, arriving on 6 May to support the landing forces there. Her powerful guns leveled the defenses at Shuri Castle, which had stalled the entire offensive. On 5 June, a kamikaze crashed into her starboard side, but the fighting ship continued to support the troops at Okinawa until 16 June.

After the announced surrender of Japan, Mississippi steamed to Sagami Wan, Honshu, arriving on 27 August as part of the support occupation force. She anchored in Tokyo Bay, witnessed the signing of the surrender documents, and steamed for home on 6 September. She arrived on 27 November at Norfolk, where she underwent conversion to AG-128, effective 10 February 1946. As part of the operational development force, she spent the last 10 years of her career carrying out investigations of gunnery problems and testing new weapons, while based at Norfolk. She helped launch the Navy into the age of the guided-missile warship when she successfully test-fired the Terrier missile on 28 January 1953 off Cape Cod. She also assisted in the final evaluation of the Petrel, a radar-homing missile, in February 1956.

Mississippi was decommissioned at Norfolk on 17 September 1956 and was sold for scrapping to the Bethlehem Steel Co. on 28 November the same year.

Mississippi received eight battle stars for World War II service.

BB-42 • USS IDAHO

(BB-42: dp. 32,000 n. 1. 624', b. 97'5", dr. 30'; s. 21 k.; cpl. 1,081; a. 12 14 ', 14 5", 4 3", 2 21" tt.; cl. New York)

The fourth ship namedIdaho (BB-42) was launched by the New York Shipbuilding Corp., Camden, N.J., on 30 June 1917, sponsored by Miss H.A. Limons, granddaughter of the Governor of Idaho, and commissioned on 24 March 1919 with Captain C.T. Vogelgesang in command.

Idaho sailed on 13 April for shakedown training out of Guantanamo Bay, and after returning to New York, received President Pessoa of Brazil for the voyage to Rio de Janeiro. Departing on 6 July with her escort, the battleship arrived in Rio on 17 July 1919. From there, she set course for the Panama Canal, arriving in Monterey, Calif., in September to join the Pacific Fleet. She joined other dreadnoughts in training exercises and reviews, including a Fleet Review by President Wilson on 13 September 1919. In 1920, the battleship carried Secretary Daniels and the Secretary of the Interior on an inspection tour of Alaska.

Upon her return from Alaska on 22 July 1920, Idaho took part in fleet maneuvers off the California coast and as far south as Chile. She continued this important training until 1925, taking part in numerous ceremonies on the West Coast during the interim. Idaho participated in the fleet review held by President Harding in Seattle shortly before his death in 1923. The battleship sailed on 15 April 1925 for Hawaii, participated in war games until 1 July, and then got underway for Samoa, Australia, and New Zealand. On the return voyage, Idaho embarked gallant Comdr. John Rodgers and his seaplane crew after their attempt to fly to Hawaii, arriving in San Francisco on 24 September 1925.

For the next six years, Idaho operated out of San Pedro on training and readiness operations off California and in the Caribbean. She sailed from San Pedro on 7 September 1931 for the East Coast, entering Norfolk Navy Yard on 30 September for modernization. The veteran battleship received better armor, "blister" antisubmarine protection, improved machinery, and tripod masts during this extensive overhaul, and was readied for many more years of useful naval service. After completion on 9 October 1934, the ship conducted shakedown in the Caribbean before returning to her home port, San Pedro, on 17 April 1935.

As war clouds gathered in the Pacific, the fleet increased the tempo of its training operations. Idaho carried out fleet tactics and gunnery exercises regularly until arriving with the battle fleet at Pearl Harbor on 1 July 1940. The ship sailed for Hampton Roads on 6 June 1941 to perform Atlantic neutrality patrol, a vital part of U.S. policy in the early days of the European fighting. She moved to Iceland in September to protect American advance bases and was on station at Hvalfjordur when the Japanese attacked Pearl Harbor on 7 December 1941, catapulting America into the war.

Idaho and sister ship Mississippi departed Iceland two days after Pearl Harbor to join the Pacific Fleet, and arrived in San Francisco via Norfolk and the Panama Canal on 31 January 1942. She conducted additional battle exercises in California waters and out of Pearl Harbor until October 1942, when she entered Puget Sound Navy Yard to be re-gunned. Upon completion of this work, Idaho again took part in battle exercises and sailed on 7 April 1943 for operations in the bleak Aleutians. There, she was the flagship of the bombardment and patrol force around Attu, where she gave gunfire support to the

Army landings on 11 May 1943. During the months that followed, she concentrated on Kiska, culminating in an assault on 15 August. The Japanese were found to have evacuated the island in late July, thus abandoning their last foothold in the Aleutians.

Idaho returned to San Francisco on 7 September 1943 to prepare for the invasion of the Gilbert Islands. Moving to Pearl Harbor, she got underway with the assault fleet on 10 November and arrived off Makin Atoll on 20 November. She supported the fighting ashore with accurate gunfire support and anti-aircraft fire, remaining in the Gilberts until sailing for Pearl Harbor on 5 December 1943.

Next on the Pacific timetable was the invasion of the Marshalls, and the veteran battleship arrived off Kwajalein early on 31 January to soften up shore positions. Again, she hurled tons of shells into Japanese positions until 5 February, when the outcome was one of certain victory. After replenishing at Majuro, she bombarded other islands in the group, then moved to Rabaul, New Ireland, for a diversionary bombardment on 20 March 1944.

Idaho returned to the New Hebrides on 25 March and, after a short stay in Australia, arrived at Kwajalein with a group of escort carriers on 8 June. From there, the ships steamed to the Marianas where Idaho began a pre-invasion bombardment of Saipan on 14 June. With this brilliantly executed landing assault underway on 15 June, the battleship moved to Guam for bombardment assignments. As the American fleet decimated Japanese carrier air power in the Battle of the Philippine Sea from 19 to 21 June, Idaho protected the precious transport area and reserve troop convoys. After returning briefly to Eniwetok from 28 June to 9 July, the ship began

pre-invasion bombardment of Guam on 12 July, and continued the devastating shelling until the main assault eight days later. As ground troops battled for the island, Idaho stood offshore providing vital fire support until anchoring at Eniwetok on 2 August 1944.

The ship continued to Espiritu Santo and entered a floating dry dock on 16 August for repairs to her "blisters." After landing rehearsals on Guadalcanal in early September, Idaho moved to Peleliu on 12 September and began bombarding the island, needed as a staging base for the invasion of the Philippines. Despite the furious bombardment, Japanese entrenchments gave assault forces stiff opposition, and the battleship remained off Peleliu until 24 September providing all-important fire support for advancing marines. She then sailed for Manus and eventually to Bremerton, Wash., where she arrived for needed repairs on 22 October 1944. This was followed by battle practice off California.

Idaho's mighty guns were needed for the next giant amphibious assault on the way to Japan. She sailed from San Diego on 20 January 1945 to join a battleship group at Pearl Harbor. After rehearsals, she steamed from the Marianas on 14 February for the invasion of Iwo Jima. As marines stormed ashore on 19 February, Idaho was again blasting enemy positions with her big guns. She remained off Iwo Jima until 7 March when she got underway for Ulithi and the last of the great Pacific assaults on Okinawa.

Idaho sailed on 21 March 1945 as part of Rear Admiral Deyo's Gunfire and Covering Group and flagship of Bombardment Unit 4. She arrived offshore on 25 March and began silencing enemy shore batteries and pounding installations. The landings began on 1 April, and as the Japanese made a desperate attempt to drive the vast fleet away with suicide attacks, Idaho's gunners shot down numerous planes. In a massed attack on 12 April, the battleship shot down five kamikazes before suffering damage to her port blisters from a near miss. After temporary repairs, she sailed on 20 April and arrived in Guam five days later.

The veteran of so many landings of the Pacific quickly completed repairs and returned to Okinawa on 22 May to resume fire support. Idaho remained until 20 June 1945, then sailed for battle maneuvers in Leyte Gulf until hostilities ceased on 16 August 1945.

Idaho made her triumphal entry into Tokyo Bay with occupation troops on 27 August and witnessed the signing of the surrender on board Missouri on 2 September. Four days later, she began the long voyage to the East Coast of the United States, steaming via the Panama Canal to Norfolk on 16 October 1945. She decommissioned on 3 July 1946 and was placed in reserve until sold for scrap on 24 November 1947 to Lipsett Inc., of New York City.

Idaho received seven battle stars for World War II service.

BB-43 • USS TENNESEE

(BB-43: displacement 33,190; length 624'; beam 97'3"; draft 31'; speed 21 knots; complement 1,401; armament 12 14-inch guns, 14 6-inch guns, 4 3-inch AA guns, 2 21-inch torpedo tubes; class Tennessee)

The fifth Tennessee (BB-43) was laid down on 14 May 1917 at the New York Navy Yard, launched on 30 April 1919, sponsored by Miss Helen Lenore Roberts, daughter of the governor of Tennessee, and commissioned on 3 June 1920 with Captain Richard H. Leigh in command.

Tennessee and her sister ship, California (BB-44), were the first American battleships built to a "post-Jutland" hull design. As a result of extensive experimentation and testing, her underwater hull protection was significantly enhanced compared to previous battleships, and both her main and secondary batteries had fire control systems. The Tennessee class, and the three ships of the Colorado class which followed, were distinguished by two heavy cage masts supporting large fire control tops. This feature distinguished the "Big Five" from the rest of the battleship force until World War II. Tennessee's 14-inch turret guns could be elevated to 30 degrees, allowing an additional range of 10,000 yards, a significant advantage as battleships began to carry airplanes for long-range gunfire spotting.

After fitting out, Tennessee conducted trials in Long Island Sound from 15 to 23 October 1920. While at New York, one of her 300-kilowatt ship's service generators exploded on 30 October, injuring two men. The ship and yard crews worked diligently to rectify the engineering issues, enabling Tennessee to depart New York on 26 February 1921 for trials at Guantanamo. She then moved north to Hampton Roads on 19 March, conducted gunnery calibration at Dahlgren, Virginia, and underwent drydocking in Boston. After full-power trials off Rockland, Maine, and a stop in New York, she transited the Panama Canal, arriving at San Pedro, California, her home port for the next 19 years, on 17 June.

Here, she joined the Battleship Force, Pacific Fleet. In 1922, this fleet was redesignated the Battle Fleet (later renamed the Battle Force in 1931), United States Fleet. For the next two decades, Tennessee served here until World War II.

Tennessee's peacetime service involved annual training, maintenance, and readiness exercises. Her schedule included gunnery and engineering competitions and annual fleet problems, large-scale war games engaging most or all of the United States Fleet in strategic and tactical situations. From Fleet Problem I in 1923 to Fleet Problem XXI in April 1940, Tennessee actively participated in these exercises. Her proficiency was demonstrated by winning the "E" for excellence in gunnery in the competitive years 1922 and 1923, and the Battle Efficiency Pennant in 1923 and 1924. In 1925, she participated in joint Army-Navy maneuvers in Hawaii and then traveled to Australia and New Zealand. Subsequent fleet problems and exercises took Tennessee from Hawaii to the Caribbean and Atlantic, and from Alaskan waters to Panama.

Fleet Problem XXI in spring 1940 was conducted in Hawaiian waters. At its conclusion, rather than returning to San Pedro, the battleship force was shifted to Pearl Harbor at President Roosevelt's direction, hoping to deter Japanese expansion in the Far East. Following an overhaul at the Puget Sound Navy Yard, Tennessee arrived at Pearl Harbor on 12 August 1940. With the worsening global situation, Fleet Problem XXII, scheduled for spring 1941, was cancelled; Tennessee's activities were limited to smaller scale operations.

On the morning of 7 December 1941, Tennessee was moored on Battleship Row at Pearl Harbor, alongside

West Virginia (BB-48). At about 0755, Japanese carrier planes began their attack. Tennessee went to general quarters, manned her anti-aircraft guns, and prepared to sortie. However, torpedo hits on Oklahoma and West Virginia trapped Tennessee in her berth. During the attack, Tennessee sustained bomb hits, one on the after mainyard and another on the barrel of a gun turret. Despite minor physical damage, Tennessee was threatened by oil fires from adjacent ships. The fires on Tennessee were brought under control by 1030, but she remained trapped for two more days.

The first priority post-attack was to free Tennessee from her berth. Demolishing the mooring quays and carefully navigating past Oklahoma's sunken hull, Tennessee moved to the Pearl Harbor Navy Yard on 16 December.

Temporary repairs were made, including rewelding and recaulking hull and weather deck seams, and patching Turret III's damaged top. Tennessee departed Pearl Harbor with Pennsylvania (BB-38) and Maryland on 20 December, arriving at the Puget Sound Navy Yard on 29 December 1941 for permanent repairs.

During early 1942, Tennessee's after hull plating and electrical wiring were repaired, her cage mainmast was replaced by a tower, and fire control radars were installed. Her anti-aircraft armament was also upgraded. She departed Puget Sound with Maryland and Colorado on 25 February 1942, arriving in San Francisco for intensive training with Rear Admiral William S. Pye's Task Force 1.

Tennessee's role in the war shifted from conventional surface battles to naval shore bombardment and gunfire support for troops, as well as patrol duty in areas where firepower was more important than speed. In June 1943,

after undergoing modernization at the Puget Sound Navy Yard, she appeared virtually transformed. New features included deep blisters for torpedo protection, rearranged internal compartmentation, a compact superstructure, and upgraded armaments, including 5"/38 twin mounts, quadruple 40-millimeter gun mounts, and 20-millimeter guns.

After training and rehearsals, Tennessee departed San Pedro on 31 May 1943 for patrol operations in the North Pacific with Task Force 16, the North Pacific Force. She participated in patrols and bombardments in the Aleutian Islands, including Kiska, during June and July 1943.

Tennessee then supported the invasions of the Gilbert and Marshall Islands, including Betio in November 1943 and Kwajalein in January 1944. During these operations, Tennessee provided intense fire support, demonstrating the effectiveness of battleship gunfire in support of amphibious assaults.

In February 1944, Tennessee participated in the capture of Eniwetok Atoll, providing bombardment support and responding to requests for illumination and fire support from ground troops. The successful capture of Eniwetok demonstrated the value of pre-landing bombardment and the role of older battleships in providing fire support for amphibious operations.

After her significant contributions in the Marshall Islands Campaign in early 1944, USS Tennessee (BB-43) continued to play a vital role in several key operations in the Pacific Theater during World War II. In the Mariana and Palau Islands campaign, including the pivotal Battle of Saipan in June 1944, Tennessee provided crucial gunfire support, aiding the invasion and suppression of enemy defenses. Her firepower also supported the successful assaults on Tinian and Guam.

During the Philippines Campaign from 1944 to 1945, Tennessee was instrumental in the liberation of the Philippines, bombarding Leyte in October 1944 to support General Douglas MacArthur's historic return. She was part of the Battle of Surigao Strait, a component of the larger Battle of Leyte Gulf, which was one of the last battleship-versus-battleship actions in naval history and a decisive victory for the Allies.

In 1945, Tennessee's guns supported the invasions of Iwo Jima and Okinawa, proving essential in softening up defenses before the landings and providing direct fire support to troops ashore. As World War II drew to a close, Tennessee continued her active engagement, including bombarding the Japanese home islands in July 1945, just before the war's end.

After Japan's surrender, Tennessee participated in Operation Magic Carpet, the extensive operation to return American servicemen from the Pacific. She was later moved to the Atlantic Fleet and placed in reserve in February 1947. Tennessee was decommissioned on 14 February 1947, remaining in the Atlantic Reserve Fleet until being struck from the Naval Vessel Register on 1 March 1959. Her journey concluded when she was sold for scrap on 10 July 1959.

BB-44 • USS CALIFORNIA

(BB-44: displacement 33,190; length 624'; beam 97'3"; draft 31'; speed 21 knots; complement 1,401; armament 12 14-inch guns, 14 6-inch guns, 4 3-inch AA guns, 2 21-inch torpedo tubes; class Tennessee)

The fifth California (BB-44) was launched on 20 November 1919 by Mare Island Navy Yard, sponsored by Mrs. R.T. Zane, and commissioned on 10 August 1921 with Captain H.J. Ziegemeier in command. It reported to the Pacific Fleet as the flagship.

From 1921 to 1941, California served as the flagship of the Pacific Fleet and later as the flagship of the Battle Fleet (Battle Force), U.S. Fleet. Its annual activities included joint Army-Navy exercises, tactical and organizational development problems, and fleet concentrations for various purposes. The ship won the Battle Efficiency Pennant for 1921-22 and the Gunnery "E" for 1925-26 due to its intensive training and superior performance.

In the summer of 1925, California led the Battle Fleet and a division of cruisers from the Scouting Fleet on a successful goodwill cruise to Australia and New Zealand. It participated in the Presidential reviews of 1927, 1930, and 1934. The ship underwent modernization in late 1929 and early 1930, receiving an improved anti-aircraft battery.

In 1940, California's base was shifted to Pearl Harbor. On 7 December 1941, during the Japanese aerial attack on Pearl Harbor, California, moored at the southernmost berth of "Battleship Row," suffered significant damage. At 0805, a bomb exploded below decks, igniting an anti-aircraft ammunition magazine and killing about 60 men. A second bomb ruptured her bow plates. Despite efforts to keep her afloat, California settled into the mud with only her superstructure remaining above the surface, resulting in 98 crew members lost and 61 wounded.

California was refloated on 26 March 1942 and dry-docked at Pearl Harbor for repairs. On 7 June, she departed under her own power for Puget Sound Navy Yard, where major reconstruction was undertaken, including improvements to protection, stability, anti-aircraft battery, and fire control system.

Departing Bremerton on 31 January 1944 for shakedown at San Pedro, California sailed from San Francisco on 5 May for the invasion of the Marianas. Off Saipan in June, she conducted effective shore bombardment and call fire missions. On 14 June, she was hit by a shell from an enemy shore battery, resulting in one fatality and nine injuries. Following the Saipan operation, her heavy guns supported assaults on Guam and Tinian (18 July - 9 August). On 24 August, she arrived at Espiritu Santo for repairs to her port bow, damaged in a collision with Tennessee (BB-43).

On 17 September 1944, California sailed to Manus to prepare for the invasion of the Philippines. From 17 October to 20 November, she played a key role in the Leyte operation, including participating in the Battle of Surigao Strait (25 October). On 1 January 1945, she departed the Palaus for the Luzon landings. While providing shore bombardment at Lingayen Gulf on 6 January, she was hit by a kamikaze plane, resulting in 44 crew members killed and 155 wounded. Despite this, she made temporary repairs on-site and continued her shore bombardment mission. She departed on 23 January for Puget Sound Navy Yard, arriving on 15 February for permanent repairs.

California returned to action at Okinawa on 15 June 1945 and remained in the area until 21 July. Two days later, she joined TF 95 to cover East China Sea minesweeping operations. After a short trip to San Pedro Bay, P.I., in August, the ship departed Okinawa on 20 September to cover the landing of the 6th Army occupation force at Wakanoura Wan, Honshu. Supporting the occupation

until 15 October, she then sailed via Singapore, Colombo, and Cape Town to Philadelphia, arriving on 7 December. She was placed in commission in reserve there on 7 August 1946, decommissioned on 14 February 1947, and sold on 10 July 1959.

BB-45 • USS COLORADO

(BB-45: dp. 32,600; 1. 624'; b. 97'6"; dr. 30'6", s. 21.17 k.; cpl. 1,080; a. 8 16", 12 5", 4 3", 4 6-pdr., 2 21" tt.;cl. Colorado)

The third ship named Colorado (BB-45) was launched on 22 March 1921 by the New York Shipbuilding Co., Camden, New Jersey, sponsored by Mrs. M. Melville, and commissioned on 30 August 1923 with Captain R.R. Belknap in command.

Colorado embarked on her maiden voyage from New York on 29 December 1923, visiting Portsmouth, England; Cherbourg and Villefranche, France; Naples, Italy; and Gibraltar before returning to New York on 14 February 1924. After repairs and final tests, she sailed for the West Coast on 11 July, arriving in San Francisco on 16 September 1924.

From 1924 to 1941, Colorado operated with the Battle Fleet in the Pacific, participating in fleet exercises and ceremonies, and occasionally returning to the East Coast for fleet problems in the Caribbean. She also cruised to Samoa, Australia, and New Zealand from 8 June to 26 September 1925 to show the flag in the Far Pacific. She assisted in earthquake relief at Long Beach, California, on 10 and 11 March 1933, and during an NROTC cruise from 11 June to 22 July 1937, she aided in the search for the missing Amelia Earhart.

Based in Pearl Harbor from 27 January 1941, Colorado operated in the Hawaiian training area, engaging in intensive exercises and war games until 25 June, when she departed for the West Coast and an overhaul at Puget Sound Navy Yard, completed on 31 March 1942.

After training on the West Coast, Colorado returned to Pearl Harbor on 14 August 1942 to finalize preparations for action. She operated near the Fiji Islands and New Hebrides from 8 November 1942 to 17 September 1943 to prevent further Japanese expansion. She departed Pearl Harbor on 21 October to provide pre-invasion bombardment and fire support for the invasion of Tarawa, returning to port on 7 December 1943. After a West Coast overhaul, Colorado returned to Lahaina Roads, Hawaiian Islands, on 21 January 1944 and departed the next day for the Marshall Islands operation, providing pre-invasion bombardment and fire support for the invasions of Kwajalein and Eniwetok until 23 February, when she headed back to Puget Sound Navy Yard for another overhaul.

Joining units bound for the Mariana Islands operation in San Francisco, Colorado sailed on 5 May 1944 via Pearl Harbor and Kwajalein for pre-invasion bombardment and fire support duties at Saipan, Guam, and Tinian from 14 June. On 24 July, during the shelling of Tinian, Colorado received 22 shell hits from shore batteries but continued supporting the invading troops until 3 August. After repairs on the West Coast, Colorado arrived in Leyte Gulf on 20 November 1944 to support American troops fighting ashore. A week later, she was hit by two kamikazes, killing 19 of her men, wounding 72, and causing moderate damage. Despite this, she bombarded Mindoro between 12 and 17 December before proceeding to Manus Island for emergency repairs. Returning to Luzon on 1 January 1945, she participated in pre-invasion bombardments in Lingayen Gulf. On 3 January, accidental gunfire hit her superstructure, killing 18 and wounding 51.

After replenishing at Ulithi, Colorado joined the pre-invasion bombardment group at Kerama Retto on 25 March 1945 for the invasion of Okinawa. She remained there, providing fire support until 22 May, when she departed for Leyte Gulf.

Returning to occupied Okinawa on 6 August 1945, Colorado sailed from there for the occupation of Japan, covering the airborne landings at Atsugi Airfield, Tokyo, on 27 August. Departing Tokyo Bay on 20 September 1945, she arrived in San Francisco on 16 October, then steamed to Seattle for the Navy Day celebration on 27 October. Assigned to "Magic Carpet" duty, she made three trips to Pearl Harbor to transport 6,357 veterans home before reporting to Bremerton Navy Yard for inactivation. She was decommissioned and placed in reserve on 7 January 1947, and sold for scrapping on 28 July 1959.

Colorado received seven battle stars for her World War II service.

BB-46 • USS MARYLAND

(BB-46: dp. 32,600; 1. 624'; b. 97'6"; dr. 30'6", s. 21.17 k.; cpl. 1,080; a. 8 16", 12 5", 4 3", 4 6-pdr., 2 21" tt.;cl. Colorado)

Maryland (BB-46) was laid down 24 April 1917 by Newport News Shipbuilding Co., Newport News, Va.; launched 20 March 1920, sponsored by Mrs. E. Brook Lee wife of the Comptroller of the State of Maryland; and commissioned 21 July 1921, Capt. C. F. Preston in command.

With a new type seaplane catapult and the first 16-inch guns mounted on a U.S. ship, Maryland was the pride of the Navy. Following an east coast shakedown she found herself in great demand for special occasions. She appeared at Annapolis for the 1922 Naval Academy graduation and at Boston for the anniversary of Bunker Hill and the Fourth of July. Between 18 August and 2., September she paid her first visit to a foreign port transporting Secretary of State Charles Evans Hughes to Rio de Janeiro for Brazil's Centennial Exposition. The next year, after fleet exercises off the Panama Canal Zone. Maryland transited the canal

in the latter part of June to join the battle fleet stationed on the west coast.

She made a good will voyage to Australia and New Zealand in 1925, and transported President-elect Herbert Hoover off the Pacific leg of his tour of Latin America in 1928. Throughout these years and the 1930's she served as a mainstay of fleet readiness through tireless training operations. In 1940 Maryland and the other battleships of the battle force changed their bases of operations to Pearl Harbor. She was present at battleship row along Ford Island when Japan struck 7 December 1941.

A gunner's mate striker, writing a letter near his machine gun, brought the first of his ship's guns into play, shooting down one of two attacking torpedo planes. Inboard of the Oklahoma and thus protected from the initial torpedo attack Maryland managed to bring all her antiaircraft batteries into action. Despite two bomb hits she continued to fire and, after the attack, sent fire fighting parties to assist her sister ships. The Japanese announced that she had been sunk, but 30 December, battered Yet sturdy, she entered the repair yard at Puget Bound Navy Yard.

She emerged 26 February.1942 not only repaired but modernized and ready for great service. During the important Battle of Midway, the old battleships, not fast enough to accompany the carriers, operated as a backup force. Thereafter Maryland engaged in almost constant training exercises until 1 August, when she returned to Pearl Harbor.

Assigned sentinel duty along the southern supply routes to Australia and the Pacific fighting fronts, Maryland and Colorado operated out of the Fiji Islands in November and advanced to the New Hebrides in February 1943. Her return to Pearl Harbor after 10 months in the heat off the South Pacific brought the installation of additional 40mm. antiaircraft protection.

In the vast amphibious campaigns of the Pacific the firepower of Maryland and her sister ships played a key role. Departing the Hawaiian Islands 20 October for the South Pacific, Maryland became flagship for Rear Adm. Harry W. Hill's Southern Attack Force in the Gilberts Invasion, with Maj. Gen. Julian C. Smith, Commander, 2d Marine Division, embarked. Early on 20 November her big guns commenced 5 days of shore bombardment and call fire assignment in support of one of the most gallant amphibious assaults in history, at Tarawa. After the island's capture, she remained in the area protecting the transports until she headed back ~ to the United States 7 December.

Maryland steamed from San Pedro 13 January 1944, rendezvoused with TF 53 at Hawaii, and sailed in time to be in position off the well-fortified Kwajelein Atoll in the Marshalls on the morning of the 31st. Assigned to reduce pillboxes and blockhouses on Roi Island, the old battleship fired splendidly all day and again the following morning until the assault waves were within 500 yards of the beach. Following the operation she steamed back to Bremerton, Wash., for new guns and an overhaul.

Two months later Maryland, again readied for battle sailed westward 5 May to participate in the biggest campaign yet attempted in the Pacific war—Saipan. Vice Adm. R. K. Turner allotted TF 52 3 days to soften up the island before the assault. Firing commenced 0545 on 14 June. Silencing

two coastal guns, Maryland encountered little opposition as she delivered one. devastating barrage after another. The Japanese attempted to strike back through the air. On the 18th the ship's guns claimed their first victim but 4 days later a Betty sneaked in flying low over the still-contested Saipan hills and found tw o anchored battleships. Crossing the bow of Pennsylvania, she dropped a torpedo which opened a gaping hole in Maryland's portside. Casualties were light and in 15 minutes she was underway for Eniwetok, and shortly thereafter to the repair yards at Pearl Harbor.

With an around-the-clock effort by the shipyard workers, on 13 August, 34 days after arrival, the ship again steamed north for the war zone. Rehearsing briefly in the Solomons she Joined Rear Adm. J. B. Oldendorf's Western Fire Support Group (TG 32.5) bound for the Palau Islands. Firing first on 12 September to cover minesweeping operations and underway After demolition teams, she continued the shore bombardment until the landing craft approached the beaches on the 15th. Four days dater organized resistance collapsed, permitting the fire support ships to retire to the Admiralty Islands.

Reassigned to the 7th Fleet, Maryland sortied 12 October to cover the important initial landings in the Philippines at Leyte. Despite floating mines, the invasion force entered Leyte Gulf! on the 18th. The bombardment the following day and the landings of the 20th went well, but the Japanese decided to contest this success with both kamikazes and a three-pronged naval attack.

Forewarned by submarines and scout planes, the American battleship-cruiser force steamed 24 October to the southern end of Leyte Gulf to protect Surigao Strait. Early on the 25th the enemy battleships Fuso and

Yamishiro led the Japanese advance into the Strait. The waiting Americans pounded the. enemy ships severely. First came torpedoes from the fleeting PT boats, then more torpedoes from this daring destroyers. Next came gunfire from the cruisers. Finally, at Q355 the readied guns of the battle. ship line opened fire. Thunderous salvos of heavy caliber fire slowed the enemy force and set the Japanese battle ships on fire. Leaving their doomed battleships behind, the decimated enemy ships fled; only a remnant of the original force escaped subsequent naval air attacks. Similarly other U.S. forces blunted and repulsed attacks by the center and northern enemy forces during the decisive Battle for Leyte Gulf.

In the aftermath of this important victory, Maryland patrolled the southern approaches to Surigao Strait until 29 October; after replenishment at Manus, Admiralties, she resumed patrol duty 10 November. Japanese air attacks continued to pose a definite threat. During a raid on 27 November, guns of TG 77.2 splashed 11 of the attacking planes. Shortly after sunset 2 days later, a determined suicide plane dove through the clouds and crashed J!Maryland between t.turrets Nos. 1 and 2. Thirty-one sailors died in the explosion and fire that followed; however, the sturdy battleship continued her

patrols until relieved 2 December. She reached Pearl Harbor 19 December and during the next 2 months workmen repaired and refitted "Fighting Mary."

After refresher training, Maryland headed for the western Pacific on 4 March 1945, arriving at Ulithi on the 16th. There, she joined Rear Adm. M. L. Deyo's TF 54 and on 21 March departed for the invasion of Okinawa. She approached the coast off Okinawa on 25 March and began pounding assigned targets along the southeastern part of the Japanese island fortress. In addition, she provided fire support during a diversionary raid on the southeast coast, drawing enemy defenses away from the main amphibious landings on the western beaches. On 3 April, she received a fire support call from Minneapolis (CA-6). The cruiser was unable to silence entrenched shore batteries with 8-inch fire and called on 'Fighting Mary's' mighty 16-inch guns for aid. The veteran battleship hurled six salvos which destroyed the enemy artillery.

Maryland continued her fire support duty until 7 April when she sailed with TF 54 to intercept a Japanese surface force to the northward. These ships, including the mighty battleship Yamato, came under intense air attacks that same day, and planes from the Fast Carrier Task

Force sank six of the 10 ships in the force. At dusk on the 7th, Maryland took her third hit from enemy planes in 10 months. A suicide plane loaded with a 500-pound bomb crashed onto the top of turret No. 3 from starboard. The explosion wiped out the 20mm mounts, causing 53 casualties. As before, however, she continued to blast enemy shore positions with devastating 16-inch fire. While guarding the western transport area on 12 April, she splashed two planes during afternoon raids.

On 14 April, Maryland left the firing line as escort for retiring transports. Steaming via the Marianas and Pearl Harbor, she reached Puget Sound on 7 May and entered the Navy Yard at Bremerton the next day for extensive overhaul. Completing repairs in August, she then entered the 'Magic Carpet' fleet. During the next four months, she made five voyages between the west coast and Pearl Harbor, returning more than 9,000 combat veterans to the United States.

Arriving in Seattle, Washington, on 17 December, she completed 'Magic Carpet' duty. She entered the Puget Sound Naval Shipyard on 15 April 1947 and was placed in commission in reserve on an inactive basis on 15 July. She decommissioned at Bremerton on 3 April 1947 and remained there as a unit of the Pacific Reserve Fleet. Maryland was sold for scrapping to Learner Co. in Oakland, California, on 8 July 1959.

On 2 June 1961, the Honorable J. Millard Tawes, Governor of Maryland, dedicated a lasting monument to the memory of the venerable battleship and her fighting men. Built of granite and bronze and incorporating the bell of 'Fighting Mary,' this monument honors a ship and her men whose service to the Nation reflected the highest traditions of the naval service. This monument is located on the grounds of the State House, Annapolis, Maryland.

Maryland received seven battle stars for World War II service.

Canceled Battleships

After World War I, the US Navy initiated an ambitions program to construct new battleships, aiming to modernize its fleet by replacing older ships. From March 1920 and April 1921, the Navy commenced construction on six new battleships belonging to the new South Dakota Class. These vessels were designed to have a displacement of 43,200 tons, and were armed twelve 16-inch guns, as their primary weaponry. These ships were to feature 13.5-inch armor plating on their sides, and equipped with four propellers, as well as four turbo-electric generators. However, the signing of the Washington Naval Treaty in February 1922 signed by the US, UK, France, Italy, and Japan, which established limits on naval tonnage, and set specific ratios for each country, [5:5:3:1.75:1.75 for the UK, the United States, Japan, Italy, and France, respectively]. As a result of the treaty, the US halted construction of all South Dakota class battleships. In addition, an earlier class battleship, the USS Washington was canceled.

Canceled battleships include:

- BB-47 ... USS Washington
- BB-49 ... USS South Dakota
- BB-50 ... USS Indiana
- BB-51 ... USS Montana
- BB-52 ... USS North Carolina
- BB-53 ... USS Iowa
- BB-54 ... USS Massachusetts

Furthermore, two other battleships, BB-65 USS Illinois and BB-66 USS Kentucky, which had been laid down early in World War II, were also canceled before completion. as it became evident during the war that aircraft carriers had become the most important capital ships.

The decision came as the strategic value of aircraft carriers overshadowed that of battleships during the conflict.

BB-48 • USS WEST VIRGINIA

(BB-48: dp. 33,590 (f.); 1. 624'0"; b. 97'335"; dr. 30'6" (mean); s. 21.0 k., cpl. 1,407; a. 8 16", 12 5", 8 3", four 6-pdrs., 2 21" tt.; cl. Colorado)

West Virginia (Battleship No. 48) was laid down on 12 April 1920 by the Newport News Shipbuilding and Drydock Co. of Newport News, VA.; reclassified to BB-48 on 17 July 1920, launched on 17 November 1921, sponsored by Miss Alice Wright Mann, daughter of Issac T. Mann, a prominent West Virginian, and commissioned on 1 December lD-23, Capt. Thomas J. Senn in command.

The most recent of the super-dreadnoughts," West Virginia embodied the latest knowledge of naval architecture; the water-tight compartmentation of her hull and her armor protection marked an advance over the design of battleships built or on the drawing boards before the Battle of Jutland.

In the months that followed, West Virginia ran her trials and shakedown and underwent post-commissioning alterations. After a brief period of work at the New York Navy Yard, the ship made the passage to Hampton Roads, although it experienced trouble with its steering gear while en route. Overhauling the troublesome gear thoroughly while in Hampton Roads, West Virginia, put to sea on the morning of 16 June 1924. At 1010, while the battleship was steaming in the center of Lynnhaven Channel, the quartermaster at the wheel reported that the rudder indicator would not answer. The ringing of the emergency bell in the steering motor room produced no response; Capt. Senn quickly ordered all engines to stop, but the engine room telegraph would not answer. It was later discovered that there was no power to the engine room telegraph or the steering telegraph.

The captain then resorted to sending orders down to the main control via the voice tube from the bridge. He ordered full speed ahead on the port engine; all stop on the starboard. Efforts continued apace over the ensuing moments to steer the ship with her engines and keep her in the channel and, when this failed, to check headway from the edge of the channel. Unfortunately, all efforts failed, and, as the ship lost headway due to an engine casualty, West Virginia grounded on the soft mud bottom. Fortunately, as Comdr. (later Admiral) Harold 1. Stark, the executive officer, reported: ". . . not the slightest damage to the hull had been sustained."

The court of inquiry investigating the grounding found that inaccurate and misleading navigational data had been supplied to the ship. The legends on the charts provided were found to have indicated uniformly greater channel width than actually existed. The court's findings thus exonerated Capt. Senn and the navigator from any blame.

After repairs had been affected, West Virginia became the flagship for the Commander, Battleship Divisions, Battle Fleet, on 30 October 1924, thus beginning her service as an integral part of the "backbone of the fleet"— as the battleships were regarded. She soon proved her worth under a succession of commanding officers—most of whom later attained flag rank. In 1925, for example, under Capt. A. J. Hepburn, the comparative newcomer to battleship ranks scored first in competitive short-range target practices. During Hepburn's tour, West Virginia garnered two trophies for attaining the highest merit in the category.

The ship later won the American Defense Cup—presented by the American Defense Society to the battleship obtaining the highest merit with all guns in short-range firing—and the Spokane Cup, presented by that city's Chamber of Commerce in recognition of the battleship's scoring the highest merit with all guns at short range. In 1925, West Virginia won the Battle Efficiency Pennant for battleships—the first time that the ship had won the coveted "Meatball." She won it again in 1927, 1932, and 1933.

During this period, West Virginia underwent a cycle of training, maintenance, and readiness exercises, taking part in engineering and gunnery competitions and the annual large-scale exercises, or "Fleet Problems." In the latter, the Fleet would be divided up into opposing sides, and a strategic or tactical situation would be played out, with the lessons learned becoming part and parcel of the

development of doctrine that would later be tested in the crucible of combat.

During 1925, the battleship took part in the joint Army-Navy maneuvers to test the defenses of the Hawaiian Islands and then cruised with the Fleet to Australia and New Zealand. In fleet exercises subsequent to the 1925 cruise, West Virginia ranged from Hawaii to the Caribbean and the Atlantic and from Alaskan waters to Panama.

In order to keep pace with technological developments in ordnance, gunnery, and fire control—as well as engineering and aviation—the ship underwent modifications designed to increase the ship's capacity to perform her designed function. Some of the alterations effected included the replacement of her initial 3-inch antiaircraft battery with 5-inch/25-caliber dual-purpose guns; the addition of platforms for .50-caliber machine guns at the foremast and maintop, and the addition of catapults on her quarterdeck, aft, and on her number III, or "high" turret.

In the closing years of the decade of the 1930s, however, it was becoming evident to many that it was only a matter of time before the United States became involved in yet another war on a grand scale. The United States Fleet thus came to be considered a grand deterrent to the country's most probable enemy— Japan. This reasoning produced the hurried dispatch of the Fleet to Pacific waters in the spring of 1939 and the retention of the Fleet in Hawaiian waters in 1940, following the conclusion of Fleet Problem XXI in April.

As the year 1941 progressed, West Virginia carried out a schedule of intensive training based on Pearl Harbor and operating in various task forces and groups in the Hawaiian operating area. This routine continued even through the unusually tense period that began in late November and extended into the next month. Such at-sea periods were usually followed by in-port upkeep, with the battleships mooring to masonry "quays" along the southeast shores of Ford Island in the center of Pearl Harbor.

On Sunday, 7 December 1941, West Virginia lay moored outboard of Tennessee (BB-43) at berth F-6 with 40 feet of water beneath her keel. Shortly before 0800, Japanese planes flying from a six-carrier task force commenced their well-planned attack on the Fleet at Pearl Harbor. West Virginia took five 18-inch aircraft torpedoes in her port side and two bomb hits— those bombs being 15-inch armor-piercing shells fitted with fins. The first bomb penetrated the superstructure deck, wrecking the port casemates and causing that deck to collapse to the level of the galley deck below.

Four casemates and the galley caught fire immediately, with the subsequent detonation of the ready-service projectiles stowed in the casemates.

The second bomb hit further aft, wrecking one Vought OS2U Kingfisher floatplane atop the "high" catapult on Turret III and pitching the second one on her top on the main deck below. The projectile penetrated the 4-inch turret roof, wrecking one gun in the turret itself. Although the bomb proved a dud, burning gasoline from the damaged aircraft caused some damage.

The torpedoes, though, ripped into the ship's port side; only prompt action by Lt. Claude V. Ricketts, the assistant fire control officer who had some knowledge of damage control techniques, saved the ship from the fate that befell Oklahoma (BB-37) moored ahead. She, too, took torpedo hits that flooded the ship and caused her to capsize.

Instances of heroic conduct on board the heavily damaged battleship proliferated in the heat of battle. The ship's commanding officer, Capt. Mervyn S. Bennion arrived on his bridge early in the battle, only to be struck down by a bomb fragment hurled in his direction when a 15-inch "bomb" hit the center gun in Tennessee's Turret II, spraying that ship's superstructure and West Virginia's with fragments. Bennion, hit in the abdomen, crumpled to the deck, mortally wounded, but clung tenaciously to life until just before the ship was abandoned, involved in the conduct of the ship's defense up to the last moment of his life. For his conspicuous devotion to duty, extraordinary courage, and complete disregard for his own life, Capt. Bennion was awarded a Medal of Honor posthumously.

West Virginia was abandoned, settling to the harbor bottom on an even keel, her fires fought from onboard by a party that volunteered to return to the ship after the first abandonment. By the afternoon of the following day, 8 December, the flames had been extinguished. The garbage lighter, YG-17, played an important role in assisting those efforts during the Pearl Harbor attack, remaining in position despite the danger posed by exploding ammunition on board the battleship.

Later examination revealed that West Virginia had taken not five but six torpedo hits. With a patch over the damaged areas of her hull, the battleship was pumped out and ultimately refloated on 17 May 1942. docked in Drydock Number One on 9 June, West Virginia again came under scrutiny, and it was discovered that there had been seven torpedo hits, not six.

During the ensuing repairs, workers located 70 bodies of West Virginia sailors who had been trapped below when the ship sank. In one compartment, a calendar was found, the last scratch-off date being 23 December. The task confronting the nucleus crew and shipyard workers was a monumental one, so great was the damage on the battleship's port side. Ultimately, however, West Virginia departed Pearl Harbor for the West Coast and a complete rebuilding at the Puget Sound Navy Yard at Bremerton, Wash.

Emerging from the extensive modernization, the battleship that had risen, Phoenix-like, from the destruction at Pearl Harbor looked totally different from the way she had appeared prior to 7 December 1941. Gone were the "cage" masts that supported the three-tier fire-control tops, as well as the two funnels, the open mount 5 inch/26 inch, and the casemates with the single-purpose 5 inch/61 inch. A streamlined superstructure now gave the ship a totally new silhouette, dual purpose 5-inch/38-caliber guns in gun houses gave the ship a potent antiaircraft battery. In addition, 40-millimeter Bofors and 20-millimeter Oerlikon batteries studded the decks, giving the ship a heavy "punch" for dealing with close-in enemy planes.

West Virginia remained at Puget Sound until early July 1944. Loading ammunition on the 2d, the battleship got underway soon thereafter to conduct her sea trials out of Port Townsend, Wash. She ran a full power trial on the 6th, continuing her working up until the 12th. Subsequently returning to Puget Sound for last-minute repairs, the battleship headed for San Pedro and her post-modernization shakedown.

In late 1944, the USS West Virginia was involved in the Leyte Gulf operations, part of the Philippines campaign, engaging in the Battle of Surigao Strait. This battle was notable as one of the last battleship versus battleship actions in naval history and a decisive American victory.

In 1945, the USS West Virginia continued its active involvement in the Pacific. It supported the landings on Iwo Jima in February and provided shore bombardment at Okinawa in April, where it remained on station for a considerable time to offer fire support and protect against potential Japanese air attacks. During the Okinawa campaign, the West Virginia was hit by a kamikaze but was able to continue its mission after quick repairs.

As the war drew to a close, the USS West Virginia participated in bombarding the Japanese home islands, contributing to the pressure on Japan prior to its surrender. Following Japan's capitulation, the battleship was present in Tokyo Bay and witnessed the formal Japanese surrender on September 2, 1945, aboard the USS Missouri.

After the war, the USS West Virginia briefly served as part of the occupation forces in Japan. The ship was eventually decommissioned in 1947 and entered the Pacific Reserve Fleet. It remained in reserve until stricken from the Naval Vessel Register in 1959, and was subsequently sold for scrap in 1959, marking the end of its service.

BB-55 • USS NORTH CAROLINA

North Carolina (BB-55: dp. 35,000; l. 728'9"; b. 108'4"; dr. 26'8"; s. 27 k.; cpl. 1,880; a. 9 16", 20 5", 16 1.1", 12 .50 cal. mg.; cl. North Carolina)

The third she named North Carolina (BB-55) was laid down on 27 October 1937 by the New York Naval Shipyard; launched on 13 June 1940; sponsored by Miss Isabel Hoey, daughter of the Governor of North Carolina; and commissioned in New York on 9 April 1941, with Captain Olaf M. Hustvedt in command.

As the first commissioned of the Navy's modern battleships, North Carolina received so much attention during her fitting out and trials that she won the enduring nickname "Showboat". North Carolina completed her shakedown in the Caribbean prior to the Pearl Harbor attack, and after intensive war exercises, entered the Pacific on 10 June 1942.

North Carolina and the Navy began the long island-hopping campaign for victory over the Japanese by landing Marines on Guadalcanal and Tulagi on 7 August 1942. After screening Enterprise (CV-6) in the Air Support Force for the invasion, North Carolina guarded the carrier during operations protecting supply and communication lines southeast of the Solomons. Enemy carriers were located on 24 August, and the Battle of the Eastern Solomons erupted. The Americans struck first, sinking carrier Ryujo; Japanese retaliation came as bombers and torpedo planes, covered by fighters, roared in on Enterprise and North Carolina. In an 8-minute action, North Carolina shot down between 7 and 14 enemy aircraft, her gunners standing to their guns despite the jarring detonation of 7 near-misses. One man was killed by a strafer, but the ship was undamaged. The protection North Carolina could offer Enterprise was limited as the speedy carrier drew ahead of her. Enterprise took three direct hits while her aircraft severely damaged seaplane carrier Chitose and hit other Japanese ships. Since the Japanese lost about 100 aircraft in this action, the United States won control of

the air and averted a threatened Japanese reinforcement of Guadalcanal.

North Carolina then gave her mighty strength to protect Saratoga (CV-3). Twice during the following weeks of support to Marines ashore on Guadalcanal, North Carolina was attacked by Japanese submarines. On 6 September, she maneuvered successfully, dodging a torpedo which passed 300 yards off the port beam. Nine days later, sailing with Hornet (CV-8), North Carolina took a torpedo portside, 20 feet below her waterline, and 5 of her men were killed. But skillful damage control by her crew and the excellence of her construction prevented disaster; a 5.5-degree list was righted in as many minutes, and she maintained her station in a formation at 25 knots.

After repairs at Pearl Harbor, North Carolina screened Enterprise and Saratoga and covered supply and troop movements in the Solomons for much of the next year. She was at Pearl Harbor in March and April 1943 to receive advanced fire control and radar gear, and again in September, to prepare for the Gilbert Islands operation.

With Enterprise, in the Northern Covering Group, North Carolina sortied from Pearl Harbor on 10 November for the assault on Makin, Tarawa, and Abemama. Air strikes began on 19 November, and for 10 days mighty air blows were struck to aid Marines ashore engaged in some of the bloodiest fighting of the Pacific War. Supporting the Gilberts campaign and preparing for the assault on the Marshalls, North Carolina's highly accurate big guns bombarded Nauru on 8 December,

destroying air facilities, beach defense revetments, and radio installations. Later that month, she protected Bunker Hill (CV-17) in strikes against shipping and airfields at Kavieng, New Ireland and in January 1944 joined Fast Carrier Striking Force 58, Rear Admiral Marc Mitscher in command, at Funafuti, Ellice Islands.

During the assault and capture of the Marshall Islands, North Carolina illustrated the classic battleship functions of World War II. She screened carriers from air attack in pre-invasion strikes as well as during close air support of troops ashore, beginning with the initial strikes on Kwajalein on 29 January. She fired on targets at Namur and Roi, where she sank a cargo ship in the lagoon. The battleship then protected carriers in the massive air strike on Truk, the Japanese fleet base in the Carolines, where 39 large ships were left sunk, burning, or uselessly beached, and 211 planes were destroyed, another 104 severely damaged. Next, she fought off an air attack against the flattops near the Marianas on 21 February, splashing an enemy plane, and the next day again guarded the carriers in air strikes on Saipan, Tinian, and Guam. During much of this period, she was flagship for Rear Admiral (later Vice Admiral) Willis A. Lee, Jr., Commander Battleships Pacific.

With Majuro as her base, North Carolina joined in the attacks on Palau and Woleai from 31 March to 1 April, shooting down another enemy plane during the approach phase. On Woleai, 150 enemy aircraft were destroyed along with ground installations. Support for the capture of the Hollandia area of New Guinea followed (13-24 April), then another major raid on Truk (29-30 April), during which North Carolina splashed yet another enemy aircraft. At Truk, North Carolina's planes were catapulted to rescue an American aviator downed off the reef. After one plane had turned over on landing and the other, having rescued all the airmen, had been unable to take off with so much weight, Tang (SS-306) saved all involved. The next day, North Carolina destroyed coast defense guns, anti-aircraft batteries, and airfields at Ponape. The battleship then sailed to repair her rudder at Pearl Harbor.

Returning to Majuro, North Carolina sortied with the Enterprise group on 6 June for the Marianas. During the assault on Saipan, North Carolina not only gave her usual protection to the carriers but starred in bombardments on the west coast of Saipan covering minesweeping operations, and blasted the harbor at Tanapag, sinking several small craft and destroying enemy ammunition, fuel, and supply dumps. At dusk on invasion day, 15 June, the battleship downed one of the only two Japanese aircraft able to penetrate the combat air patrol.

On 18 June, North Carolina cleared the islands with the carriers to confront the Japanese 1st Mobile Fleet, tracked by submarines and aircraft for the previous four days. The next day began the Battle of the Philippine Sea, and she took station in the battle line that fanned out from the carriers. American aircraft succeeded in downing most of the Japanese raiders before they reached the American ships, and North Carolina shot down two of the few which got through.

On that day and the next, American air and submarine attacks, with the fierce anti-aircraft fire of such ships as North Carolina, virtually ended any future threat from Japanese naval aviation: three carriers were sunk, two tankers damaged so badly they were scuttled, and all but 35 of the 430 planes with which the Japanese had begun the battle were destroyed. The loss of trained aviators was irreparable, as was the loss of skilled aviation maintenance men in the carriers. Not one American ship was lost, and only a handful of American planes failed to return to their carriers.

After supporting air operations in the Marianas for another two weeks, North Carolina sailed for overhaul at Puget Sound Navy Yard. She rejoined the carriers off Ulithi on 7 November as a furious typhoon struck the group. The ships fought through the storm, and carried out air strikes against western Leyte, Luzon, and the Visayas to support the struggle for Leyte. During similar strikes later in the month, North Carolina fought off her first kamikaze attack.

As the pace of operations in the Philippines intensified, North Carolina guarded carriers while their planes kept the Japanese aircraft on Luzon airfields from interfering with the invasion convoys which assaulted Mindoro on 15 December. Three days later the task force again sailed through a violent typhoon, which capsized several

destroyers. With Ulithi now her base, North Carolina screened wide-ranging carrier strikes on Formosa, the coast of Indo-China and China, and the Ryukyus in January, and similarly supported strikes on Honshu the next month. Hundreds of enemy aircraft were destroyed which might otherwise have resisted the assault on Iwo Jima, where North Carolina bombarded and provided call fire for the assaulting Marines through 22 February.

Strikes on targets in the Japanese home islands laid the groundwork for the Okinawa assault, in which North Carolina played her dual role of bombardment and carrier screening. Here, on 6 April, she downed three kamikazes, but took a 5-inch hit from a friendly ship during the melee of antiaircraft fire. Three men were killed and 44 wounded. The next day came the last desperate sortie of the Japanese Fleet, as Yamato, the largest battleship in the world, came south with her attendants. Yamato, a cruiser, and a destroyer were sunk, three other destroyers damaged so badly that they were scuttled and the remaining four destroyers returned to the fleet base at Sasebo badly damaged. On the same day, North Carolina splashed an enemy plane, and she shot down two more on 17 April.

After overhaul at Pearl Harbor, North Carolina rejoined the carriers for a month of air strikes and naval bombardment on the Japanese home islands. Along with guarding the carriers, North Carolina fired on major industrial plants near Tokyo, and her scout plane pilots performed a daring rescue of a downed carrier pilot under heavy fire in Tokyo Bay.

North Carolina sent both sailors and members of her Marine Detachment ashore for preliminary occupation duty in Japan immediately at the close of the war, and patrolled off the coast until anchoring in Tokyo Bay on 5 September to reembark her men. Carrying passengers from Okinawa, North Carolina sailed for home, reaching the Panama Canal on 8 October. She anchored at Boston on 17 October, and after overhaul at New York, exercised in New England waters and carried Naval Academy midshipmen for a summer training cruise in the Caribbean.

After inactivation, she was decommissioned at New York on 27 June 1947. Struck from the Navy List on 1 June 1960, North Carolina was transferred to the people of North Carolina on 6 September 1961. On 29 April 1962, she was dedicated at Wilmington, N.C., as a memorial to North Carolinians of all services killed in World War II. Here, splendidly maintained and most appropriately displayed — including a spectacular "sound and light" presentation — "Showboat" still serves mightily to strengthen and inspire the nation.

North Carolina received 12 battle stars for World War II service.

BB-56 • USS WASHINGTON

(BB-56: dp. 35,000; 1. 729', b. 108' dr. 38', s. 27 k. Cpl. 1,880; a. 9 16", 20 5", 16 1.1" mg.; cl. North Carolina)

The eighth ship named Washington (BB-56) was laid down on 14 June 1938 at the Philadelphia Navy Yard, launched on 1 June 1940; sponsored by Miss Virginia Marshall, of Spokane, Wash., a direct descendant of former Chief Justice Marshall; and commissioned at the Philadelphia Navy Yard on 15 May 1941, Capt. Howard H. J. Benson in command.

Her shakedown and underway training ranged along the eastern seaboard and into the Gulf of Mexico and lasted through American entry into World War II in December 1941. Sometimes operating in company with her sistership North Carolina (BB-55) and the new aircraft carrier Hornet (CV-8), Washington became the flagship for Rear Admiral John W. Wilcox, Commander, Battleship Division (ComBatDiv) 6, and Commander, Battleships, Atlantic Fleet.

Assigned duty as flagship for Task Force (TF) 39 on 26 March 1942 at Portland, Maine, Washington again flew Admiral Wilcox's flag as she sailed for the British Isles that day. Slated to reinforce the British Home Fleet, the battleship, together with the carrier Wasp (CV-78) and the heavy cruisers Wichita (CA-45) and Tuscaloosa (CA 37), headed for Scapa Flow, the major British fleet base in the Orkney Islands.

While steaming through moderately heavy seas the following day, 27 March, the "man overboard" alarm sounded on board Washington, and a quick muster revealed that Admiral Wilcox was missing. Tuscaloosa, 1,000 yards astern, maneuvered and dropped life buoys while two destroyers headed for Washington's wake to search for the missing flag officer. Planes from Wasp, despite the foul weather, also took off to aid in the search.

Lookouts in the destroyer Wilson (DD-408) spotted Wilcox's body in the water, face down, some distance away, but could not pick it up. The circumstances surrounding

Wilcox being washed overboard from his flagship have never been fully explained to this day. One school of thought has it that he had suffered a heart attack.

At 1228 on the 27th, the search for Wilcox was abandoned, and command of the task force devolved upon the next senior officer, Rear Admiral Robert C. Giffen, whose flag flew in the cruiser Wichita. On 4 April, the task force reached Scapa Flow, joining the British Home Fleet under the overall command of Sir John Tovey, whose flag flew on the battleship HMS King George V.

Washington engaged in maneuvers and battle practice with units of the Home Fleet, out of Scapa Flow, into late April, when TF 39 was redesignated as TF 99 with Washington as flagship. On the 28th, the force got underway to engage in reconnaissance for the protection of the vital convoys running lend-lease supplies to Murmansk in the Soviet Union.

During those operations, tragedy befell the group. On 1 May 1942, HMS King George V collided with a "Tribal"-class destroyer. HMS Punjab, cut in two, sank quickly directly in the path of the oncoming Washington. Compelled to pass between the halves of the sinking destroyer, the battleship proceeded ahead, Punjabi's depth charges exploding beneath her hull as she passed.

Fortunately for Washington, she suffered no major hull damage nor developed any hull leaks from the concussion of the exploding depth charges. She did, however, sustain damage to some of her delicate fire control systems and radars, and a diesel oil tank suffered a small leak.

Meanwhile, two destroyers picked up Punjabi's captain, four other officers, and 182 men; HMS King George V then proceeded back to Scapa Flow for repairs. Washington and her escorts remained at sea until 6 May, when TF 99 put into the Icelandic port of Hvalfjordur to provision from the supply ship Mizar (AF-12). While at Hvalfjordur, the American and Danish ministers to Iceland called upon Admiral Giffen and inspected his flagship on 12 May.

Task Force 99 subsequently sortied on the 15th to rendezvous with units of the Home Fleet and returned to Scapa Flow on 3 June. The next day, Admiral Harold

R. Stark, Commander of Naval Forces, Europe, came on board and broke his flag in Washington, establishing a temporary administrative headquarters on board. The battleship played host to His Majesty, King George VI, at Scapa Flow on the 7th when the King came on board to inspect the ship.

Soon after Admiral Stark left Washington, the battleship resumed her operations with the Home Fleet patrolling part of the Allied shipping lanes leading to Russian ports. On 14 July 1942, Admiral Giffen hauled down his flag in the battleship at Hvalfjordur and shifted to Wichita. That same day, Washington, with a screen of four destroyers, upped-anchor and put to sea, leaving Icelandic waters in her wake. She reached Gravesend Bay, N.Y., on 21 July; two days later, she shifted to the New York Navy Yard, Brooklyn, N.Y., for a thorough overhaul.

Upon completion of her refit, Washington sailed for the Pacific on 23 August, escorted by three destroyers. Five days later, she transited the Panama Canal and, on 14 September, reached Nukualofa Anchorage Tongatabu, Tonga Island. On that day, Rear Admiral Willis A. "Ching" Lee, Jr., broke his flag in Washington as Commander of Battleship Division (BatDiv) 6 and Commander, Task Group 12.2.

The next day, 15 September, Washington put to sea bound for a rendezvous with TF 17, the force formed around the aircraft carrier Hornet. Washington then proceeded to Noumea, New Caledonia, and supported the ongoing Solomons campaign, providing escort services for various reinforcement convoys proceeding to and from Guadalcanal. During those weeks, the battleship's principal bases of operation were Noumea and Espiritu Santo, New Hebrides.

By mid-November, the situation in the Solomons was far from good for the Allies, who were now down to one aircraft carrier—Enterprise (CV-6)—after the loss of Wasp in September and Hornet in October, and Japanese surface units were subjecting Henderson Field on Guadalcanal to heavy bombardments with disturbing regularity. Significantly, however, the Japanese only made their moves at night since Allied planes controlled the skies during the day. That meant that the Allies had to move their replenishment and reinforcement convoys into Guadalcanal during the daylight hours.

Washington performed those vital duties into mid-November of 1942. On 13 November, she learned that three groups of Japanese ships—one consisting of about 24 transports, with escort—were steaming toward Guadalcanal. One enemy force sighted that morning was reported to consist of two battleships, a light cruiser, and 11 destroyers.

At sunset on the 13th, Rear Admiral Lee took Washington, South Dakota (BB-57), and four destroyers and headed for Savo Island—the scene of the disastrous night action of 8 and 9 August—to be in position to intercept the Japanese convoy and its covering force. Lee's ships, designated as TF 64, reached a point about 50 miles south-by-west from Guadalcanal late in the forenoon on the 14th and spent much of the remainder of the day trying—unsuccessfully—to avoid being spotted by Japanese reconnaissance planes.

Approaching a northerly course nine miles west of Guadalcanal, TF 64—reported by the Japanese reconnaissance planes as consisting of a battleship, a cruiser, and four destroyers—steamed in column formation. Walke (DD-416) led, followed by Benham (DD-397), Preston (DD-377), Gwin (DD-433), and the two battleships, Washington and South Dakota.

As the ship steamed through the flat, calm sea beneath the scattered cirrus cumulus clouds in the night sky, Washington's radar picked up a contact, bearing to the east of Savo Island, at 0001 on 15 November. Fifteen minutes later, at 0016, Washington opened fire with her 16-inch main battery. The fourth battle of Savo Island was underway.

The Japanese force proved to be the battleship Kirishima, the heavy cruisers Atago and Takao, the light cruisers Sendai and Nagara, and a screen of nine destroyers escorting four transports. Planning to conduct a bombardment of American positions on Guadalcanal to cover the landing of troops, the Japanese force ran head-on into Lee's TF 64.

For the next three minutes, Washington's 16-inchers hurled out 42 rounds, opening at 18,500 yards range, her

fire aimed at the light cruiser Sendai. Simultaneously, the battleship's 5-inch battery was engaging another ship that was also being engaged by South Dakota.

As gun-flashes split the night and the rumble of gunfire reverberated like thunder off the islands nearby, Washington continued to engage the Japanese force. Between 0025 and 0034, the ship engaged targets at 10,000 yards range with her 5-inch battery.

Most significantly, however, Washington soon engaged Kirishima in the first head-to-head confrontation of battleships in the Pacific War. In seven minutes of tracking by radar, Washington sent 75 rounds of 16-inch and 107 rounds of 5-inch at ranges from 8,400 to 12,650 yards, scoring at least nine hits with her main battery and about 40 with her 5-inches, silencing the enemy battleship in short order. Subsequently, Washington's 5-inch batteries went to work on other targets spotted by her radar "eyes."

The battle, however, was not all one-sided. Japanese gunfire proved devastating to the four destroyers of TF 64, as did the dreaded and effective "long lance" torpedoes. Walke and Preston both took numerous hits of all calibers and sank; Benham sustained heavy damage to her bow, and Gwin sustained shell hits aft.

South Dakota had maneuvered to avoid the burning of Walke and Preston but soon found herself the target of the entire Japanese bombardment group. Skewered by searchlight beams, South Dakota boomed out salvoes at the pugnacious enemy, as did Washington, which was proceeding, at that point, to deal out severe punishment upon Kirishima—one of South Dakota's assailants.

South Dakota, the recipient of numerous hits, retired as Washington steamed north to draw fire away from her crippled sister battleship and the two crippled destroyers, Benham and Gwin. Initially, the remaining ships of the Japanese bombardment group gave chase to Washington but broke off action when discouraged by the battleship's heavy guns. Accordingly, they withdrew under cover of a smokescreen.

After Washington skillfully evaded torpedoes fired by', the retiring Japanese destroyers in the van of the enemy force, she joined South Dakota later in the morning, shaping course for Noumea. In the battleship action, Washington had done well and had emerged undamaged. South Dakota had not emerged unscathed, however, sustaining heavy damage to her superstructure; 38 men had died, and 60 lay wounded. The Japanese had lost the battleship Kirishima. Left burning and exploding, she later had to be abandoned and scuttled. The other enemy casualty was the destroyer Ayanami, who was scuttled the next morning.

Washington remained in the South Pacific theater, basing on New Caledonia and continuing as flagship for Rear Admiral "Ching" Lee. The battleship protected carrier groups and task forces engaged in the ongoing Solomons campaign until late in April of 1943, operating principally with TF 11, which included the repaired Saratoga (CV-3), and with TF 16, built around Enterprise.

Washington departed Noumea on 30 April 1943, bound for the Hawaiian Islands. While en route, TF 16 joined up, and together, the ships reached Pearl Harbor on 8 May. Washington, as a unit of, and as flagship for, TF 60, carried out battle practice in Hawaiian waters until 28 May 1943, after which time she was put into the Pearl Harbor Navy Yard for overhaul.

Washington resumed battle practice in the Hawaiian operating area upon the conclusion of those repairs and alterations and joined a convoy on 27 July to form Task Group (TG) 56.14, bound for the South Pacific. Detached on 5 August, Washington reached Havannah Harbor, at Efate, in the New Hebrides, on the 7th. She operated out of Efate until late October, principally engaged in battle practice and tactics with fast carrier task forces.

Departing Havannah Harbor on the last day of October, Washington sailed as a unit of TG 53.2— four battleships and six destroyers. The next day, the carriers Enterprise, Essex (CV-9), and Independence (CVL-22), as well as the other screening units of TG 53.3, joined TG 53.2 and came under Rear Admiral Lee. The ships held combined maneuvers until 5 November, when the carriers departed the formation. Washington, with her escorts, steamed to Viti Levu in the Fiji Islands, arriving on the 7th.

Four days later, however, the battleship was again underway, with Rear Admiral Lee—by that point Commander, Battleships, Pacific—embarking in company with other units of BatDivs 8 and 9. On the 15th, the battlewagons and their screens joined Rear Admiral C. A. "Baldy" Pownall's TG 50.1, Rear Admiral Pownall flying his two-starred flag in Yorktown (CV-10), the namesake of the carrier lost at Midway. The combined force then proceeded toward the Gilbert Islands to join in the daily bombings of Japanese positions in the Gilberts and Marshalls—softening them up for impending assault.

On the 19th, the planes from TG 50.1 attacked Mili and Jaluit in the Marshalls, continuing those strikes through 20 November, the day upon which Navy, Marine, and Army forces landed on Tarawa and Makin in the Gilberts. On the 22nd, the task group sent its planes against Mili in successive waves. Subsequently, the group steamed to operate north of Makin.

Washington rendezvoused with other carrier groups that composed TF 50 on 25 November and, during the reorganization that followed, was assigned to TG 50.4, the fast carrier task group under the command of Rear Admiral Frederick C. "Ted" Sherman. The carriers comprising the core of the group were Bunker Hill (CV-17) and Monterey (CVL-26); the battleships screening them were Alabama (BH-50) and South Dakota. Eight destroyers rounded out the screen.

The group operated north of Makin, providing air, surface, and antisubmarine protection for the unfolding unloading operations at Makin, effective on 26 November. Enemy planes attacked the group on the 27th and 28th but were driven off without inflicting any damage on the fast carrier task forces.

As the Gilbert Islands campaign drew to a close, TG 50.8 was formed on 6 December under Rear Admiral Lee in Washington. Other ships of that group included sister ship North Carolina (BB-55), Massachusetts (BB-59), Indiana (BB-58), South Dakota, and Alabama (BB-60) and the Fleet carriers Bunker Hill and Monterey. Eleven destroyers screened the heavy ships.

The group first steamed south and west of Ocean Island to take position for the scheduled air and surface bombardment of the island of Nauru. Before dawn on 8 December, the carriers launched their strike groups while the bombardment force formed in the column. One hundred thirty-five rounds of 16-inch fire from the six battleships fell on the enemy installations on Nauru, and, upon completion of the shelling, the battleships' secondary batteries took their turn; two planes from each battleship spotted the fall of shot.

After a further period of air strikes had been flown off against Nauru, the task group sailed for Efate, where they arrived on 12 December. On that day, due to a change in the highest command echelons, TF 57 became TF 37.

Washington tarried at Efate for less than two weeks. Underway on Christmas Day, flying Rear Admiral Lee's flag, the battleship sailed in company with her sistership North Carolina and a screen of four destroyers to conduct gunnery practice, returning to the New Hebrides on 7 January 1944.

Eleven days later, the battleship departed Efate for the Ellice Islands. Joining lG 37.2—carriers Monterey and Bunker Hill and four destroyers—en route, Washington reached Funafuti, Ellice Islands, on 20 January. Three days later, the battleship, along with the rest of the task group, was put to sea to make rendezvous with elements of TF

58, the fast carrier task force under the overall command of Vice Admiral Marc A. "Pete" Mitscher. Becoming part of TG 58.1, Washington screened the fast carriers in her group as they launched air strikes on Taroa and Kwajalein in the waning days of January 1944. Washington, together with Massachusetts and Indiana, left the formation with four destroyers as screen and shelled Kwajalein Atoll on the 30th. Further air strikes followed the next day.

On 1 February, however, misfortune reared her head; Washington, while maneuvering in the inky darkness, rammed Indiana as she cut across Washington's bow and dropped out of formation to fuel escorting destroyers. Both battleships retired for repairs, Washington having sustained 60 feet of crumpled bow plating. Both ships were put into the lagoon at Majuro the next morning. Subsequently, after reinforcing the damaged bow, Washington departed Majuro on 11 February, bound for the Hawaiian Islands.

With a temporary bow fitted at the Pearl Harbor Navy Yard, Washington continued on to the west coast of the United States. Reaching the Puget Sound Navy Yard, Bremerton, Wash., the battleship received a new bow over the weeks that followed her arrival. Joining BatDiv 4 at Port Townsend, Wash., Washington embarked 600 men as passengers and sailed for Pearl.

Harbor, reaching her destination on 13 June and disembarking her passengers.

Arriving back at Majuro on 30 May, Washington again flew Admiral Lee's flag as he shifted onboard the battleship soon after her arrival. Lee, now a vice admiral, rode in the

battleship as she headed out to sea again, departing Majuro on 7 June and joining Mitscher's fast carrier TP 58.

Washington supported the air strikes pummeling enemy defenses in the Marianas on the islands of Saipan, Tinian, Guam, Rota, and Pagan. Task Force 58's fliers also attacked twice and damaged a Japanese convoy in the vicinity on 12 June. The following day, Vice Admiral Lee's battleship-destroyer task group was detached from the main body of the force and conducted shore bombardment against enemy installations on Saipan and Tinian. Relieved on the 14th by two task groups under Rear Admirals J. B. Oldendorf and W. L. Ainsworth, Vice Admiral Lee's group retired momentarily.

On 15 June, Admiral Mitscher's TF 58 planes bombed Japanese installations on Iwo Jima in the Volcano Islands and Chichi Jima and Haha Jima in the Bonins. Meanwhile, marines landed on Saipan under cover of intensive naval gunfire and carrier-based planes.

That same day, Admiral Jisaburo Ozawa, commanding the main body of the Japanese Fleet, was ordered to attack and destroy the invasion force in the Marianas. The departure of his carrier group, however, came under the scrutiny of the submarine Redfin (SS-272) as it left Tawi Tawi, the westernmost island in the Sulu Archipelago.

Flying Fish (SS-229) also sighted Ozawa's force as it entered the Philippine Sea. Cavalla (SS-244) radioed a contact report on an enemy refueling group on 16 June and continued tracking it as it headed for the Marianas. She again sighted Japanese Combined Fleet units on 18 June.

Admiral Raymond A. Spruance, commanding the 5th Fleet, had meanwhile learned of the Japanese movement and accordingly issued his battle plan. Vice Admiral Lee's force formed a protective screen around the vital fleet carriers. Washington, six other battleships, four heavy cruisers, and 14 destroyers were deployed to cover the flattops. On 19 June, the ships came under attack from Japanese carrier-based and land-based planes as the Battle of the Philippine Sea commenced.

The tremendous firepower of the screen, however, together with the aggressive combat air patrols flown from the American carriers, proved to be too much for even the aggressive Japanese. The heavy loss of Japanese aircraft sometimes referred to as the "Marianas Turkey Shoot," caused serious losses in the Japanese naval air arm. During four massive raids, the enemy launched 373 planes—only 130 returned.

In addition, 50 land-based bombers from Guam fell in flames. Over 300 American carrier planes were involved in the aerial action; their losses amounted to comparatively few: 23 were shot down, and six lost operationally without the loss of a single ship in Mitscher's task force.

Only a few of the enemy planes managed to get through the barrage of flak and fighters, one scoring a direct hit on South Dakota—killing 27 and wounding 23. A bomb burst over the flight deck of the carrier Wasp (CV-18), killing one man, wounding 12, and covering her flight deck with bits of phosphorus. Two planes dove on Bunker Hill, one scoring a near miss and the other a hit that holed an elevator, knocking out the hanger deck gasoline system temporarily, killing three and wounding 73. Several fires started were promptly quenched. In addition, Minneapolis (CA-36) and Indiana also received slight damage.

Not only did the Japanese lose heavily in planes, but two of their carriers were soon on their way to the bottom—Taiho, torpedoed and sunk by Albacore (SS-218), and Shokaku, sunk by Cavalla. Admiral Ozawa, his flagship, Taiho, sunk out from under him and transferred his flag to Zuikaku.

As the Battle of the Philippine Sea proceeded to a close, the Japanese Mobile Fleet steamed back to its bases, defeated. Admiral Mitscher's task force meanwhile retired to cover the invasion operations proceeding in the Marianas. Washington fueled east of that chain of islands and then continued her screening duties with TG 58.4 to the south and west of Saipan, supporting the continuing air strikes on islands in the Marianas; the strikes concentrated on Guam by that point.

On 25 July, aircraft of TG 58.4 conducted air strikes on the Palaus and on enemy shipping in the vicinity, continuing their schedule of strikes through 6 August. On that day, Washington, with Iowa (BB-61), Indiana, Alabama, the light cruiser Birmingham (CL-62), and a destroyer screen was detached from the screen of TG 58.4, forming TG 58.7 under Vice Admiral Lee.

That group arrived at Eniwetok Atoll in the Marshalls to refuel and replenish on 11 August and remained there for almost the balance of the month. On 30 August, that group departed, headed for, first, the Admiralty Islands and, ultimately, the Palaus.

Washington's heavy guns supported the taking of Peleliu and Angaur in the Palaus and supported the carrier strikes on Okinawa on 10 October, on northern Luzon and Formosa from 11 to 14 October, as well as the Visayan air strikes on 21 October. From 5 November 1944 to 17 February 1945, Washington, as a vital unit of the fast carrier striking forces, supported raids on Okinawa, in the Ryukyus; Formosa; Luzon; Camranh Bay, French Indochina, Saigon, French Indochina; Hong- Kong, Canton, Hainan Island; Nansei Shoto, and the heart of the enemy homeland—Tokyo itself.

From 19 to 22 February 1945, Washington's heavy rifles hurled 16-inch shells shoreward in support of the landings on Iwo Jima. In preparation for the assault, Washington's main and secondary batteries destroyed gun positions, troop concentrations, and other ground installations. From 23 February to 16 March, the fast battleship supported the unfolding invasion of Iwo Jima, including a carrier raid upon Tokyo on 25 February. On 18, 19, and 29 March, Washington screened the Fleet's carriers as they launched airstrikes against Japanese airfields and other installations on the island of Kyushu. On 24 March and again on 19 April, Washington lent her support to the shellings of Japanese positions on the island of Okinawa.

Anchoring at San Pedro Bay, Leyte, on 1 June 1945 after an almost ceaseless slate of operations, Washington sailed for the west coast of the United States on 6 June, making stops at Guam and Pearl Harbor before reaching the Puget Sound Navy Yard on 23 June.

As it turned out, Washington would not participate in active combat in the Pacific theater again. Her final wartime refit carried on through V-J Day in mid-August of 1945 and the formal Japanese surrender in Tokyo Bay on 2 September. She completed her post-repair trials and conducted underway training out of San Pedro, Calif., before she headed for the Panama Canal, returning to the Atlantic Ocean. Joining TG 11.6 on 6 October, with Vice Admiral Frederick C. Sherman in overall command, she soon transited the Panama Canal and headed for Philadelphia, the place where she had been "born." Arriving at the Philadelphia Naval Shipyard on 17 October, she participated in Navy Day ceremonies there on the 27th.

Assigned to troop transport duty on 2 November 1945—as part of the "Magic Carpet" operations—Washington went into dockyard hands on that day, emerging on the 15th with additional bunking facilities below and a crew that now consisted of only 84 officers and 835 men. Sailing on 15 November for the British Isles, Washington reached Southampton, England, on 22 November.

After embarking on a voyage with 185 army officers and 1,479 enlisted men, Washington sailed for New York. She completed that voyage and, after that brief stint as a transport, was placed out of commission, in reserve, on 27 June 1947. Assigned to the New York group of the Atlantic Reserve Fleet, Washington remained inactive through the late 1950s, ultimately being struck from the Navy list on 1 June 1960. The old warrior was sold on 24 May 1961 to the Lipsett Division, Luria Bros., of New York City, and was scrapped soon thereafter.

Washington (BB-56) earned 13 battle stars during World War II in operations that had carried her from the Arctic Circle to the western Pacific.

BB-57 • USS SOUTH DAKOTA

(BB-57: displacement 35,000; length 680'; beam 108'2½"; draft 36'4½"; speed 27.8 knots; complement 2,354; armament 9 16-inch guns, 16 5-inch guns, 68 40mm guns, 76 20mm guns; class South Dakota)

The second South Dakota (BB-57) was laid down on 5 July 1939 at Camden, N.J., by the New York Shipbuilding Corp., launched on 7 June 1941, sponsored by Mrs. Harlan J. Bushfield, and commissioned on 20 March 1942 with Captain Thomas L. Gatch in command.

After fitting out in Philadelphia, South Dakota conducted shakedown training from 3 June to 26 July. She departed the Philadelphia Navy Yard on 16 August for Panama, transiting the Panama Canal on 21 August and setting course for the Tonga Islands, arriving at Nukualofa, Tongatabu, on 4 September. Two days later, she struck an uncharted coral pinnacle in Lahai Passage, sustaining extensive hull damage. On 12 September, she sailed for the Pearl Harbor Navy Yard for repairs.

South Dakota was ready for sea again on 12 October and began training with Task Force (TF) 16, centered around aircraft carrier Enterprise (CV-6). The task force departed from Pearl Harbor on 16 October to join TF 17, centered on carrier Hornet (CV-8), northeast of Espiritu Santo. They rendezvoused on the 24th, forming TF 61 under Rear Admiral T.C. Kinkaid, ordered to sweep the Santa Cruz Islands and then move southwest to intercept Japanese forces approaching Guadalcanal.

"Catalina" patrol bombers spotted a Japanese carrier force on the 25th, prompting TF 61 to steam northwest for interception. Early the next morning, a Japanese scout triggered the Battle of Santa Cruz. During the battle, South Dakota, operating near Enterprise, provided antiaircraft fire and sustained a 500-pound bomb hit on her number one turret. The American forces retired towards Noumea, New Caledonia, that evening, with South Dakota credited with downing 26 enemy planes.

On 30 October, South Dakota and Mahan (DD-364) collided while avoiding a submarine, damaging both ships. Both continued to Noumea, where Vestal (AR-4) repaired South Dakota's collision and battle damage.

On 11 November, South Dakota, part of TF 16, sortied from Noumea for Guadalcanal. On 13 November, she joined battleship Washington (BB-56) and several destroyers to form TF 64 under Rear Admiral W.A. Lee. On the evening of the 14th, they engaged Admiral Kondo's bombardment group, including battleship Kirishima, heavy cruisers Takao and Atago, and a destroyer screen. In the battle, South Dakota sustained 42 hits, causing significant damage but remained operational.

Prometheus (AR-3) performed temporary repairs at Noumea, allowing South Dakota to sail on the 25th for Tongatabu and then to New York for overhaul, arriving on 18 December 1942.

After returning to sea on 25 February 1943, she operated with Ranger (CV-4) in the North Atlantic until mid-April, then joined the British Home Fleet at Scapa Flow until 1 August, returning to Norfolk. On 21 August, she headed for Efate Island, arriving at Havannah Harbor on 14 September.

She moved to Fiji on 7 November and sortied with Battleship Divisions 8 and 9 in support of Operation "Galvanic," the Gilbert Islands assault.

South Dakota participated in various operations, including bombardments, throughout 1943 and 1944, supporting carrier strikes and engaging in surface action. On 7 April 1945, she was part of the force that sank Japan's battleship Yamato.

On 6 May, an explosion in a powder tank caused a fire and additional explosions on South Dakota, resulting in casualties and damage. After repairs in Guam, she resumed operations, bombarding Japanese positions in July and August.

After Japan's surrender, South Dakota entered Tokyo Bay on 29 August. She departed Tokyo Bay on 20 September, proceeding via Okinawa and Pearl Harbor to the U.S. West Coast. After a brief stay in San Francisco and San Pedro, she sailed to Philadelphia for overhaul in January 1946. She was attached to the Atlantic Reserve Fleet in June and placed in reserve, out of commission, on 31 January 1947. The battleship remained in this status until struck from the Navy list on 1 June 1962. On 25 October 1962, she was sold to Lipsett Division, Luria Bros. & Co., Inc., for scrap.

BB-58 • USS INDIANA

(BB-58: dp. 36,000; l. ff80'; b. 108'2"; dr. 29'3"; s. 27 k.; cpl. 2,600; a. 9 16", 20 6", 24 40mm., 16 20mm., cl. Indiana)

Indiana (BB-58) was launched by Newport News Shipbuilding & Dry Dock Co., Newport News, Va., on 21 November 1941; sponsored by Mrs. Lewis C. Robbins, daughter of Indiana governor Henry F. Schricker; and commissioned on 30 April 1942, with Captain A. S. Merrill in command.

Following shakedown in Casco Bay, Maine, the new battleship steamed through the Panama Canal to bolster U.S. fleet units in the Pacific during the critical early months of World War II. She joined Rear Admiral Lee's carrier screening force on 28 November 1942. For the next 11 months, Indiana helped protect carriers Enterprise and Saratoga, supporting American advances in the Solomons.

Indiana steamed to Pearl Harbor on 21 October 1943, and departed on 11 November with the support forces designated for the invasion of the Gilbert Islands.

The battleship protected the carriers which supported the Marines during the bloody fight for Tarawa. Then, in late January 1944, she bombarded Kwajalein for eight days prior to the Marshall Island landings on 1 February. While maneuvering to refuel destroyers that night, Indiana collided with battleship Washington. Temporary repairs to her starboard side were made at Majuro, and she arrived at Pearl Harbor on 13 February for additional work.

Indiana joined the famed Task Force 58 for the Truk raid on 29-30 April and bombarded Ponape Island on 1 May. In June, the battlewagon proceeded to the Marianas with a giant American fleet for the invasion of that strategic group. She bombarded Saipan on 13-14 June and brought down several enemy aircraft while fighting off concentrated air attacks on 15 June. As the Japanese fleet closed on the Marianas for a decisive naval battle, Indiana steamed out to meet them as part of Rear Admiral Lee's battle line. The great fleets approached each other on 19 June for the biggest carrier engagement of the war, and as

four large air raids hit the American formations, Indiana, aided by other ships in the screens and carrier planes, downed hundreds of the attackers. With able assistance from submarines, Mitscher sank two Japanese carriers in addition to inflicting fatal losses on the enemy naval air arm during "The Great Marianas Turkey Shoot."

Indiana shot down several planes and sustained only two near torpedo misses. The issue decided, the battleship resumed her screening duties around the carriers and stayed at sea for 84 days in daily support of the Marianas invasion.

In August, the battleship began operations as a unit of Task Group 38.3, bombarding the Palaus, and later the Philippines. She screened strikes on enemy shore installations from 12-20 September 1944, helping to prepare for the coming invasion of Leyte. Indiana departed for Bremerton, Wash., arriving on 23 October.

Reaching Pearl Harbor on 12 December, the battleship immediately began underway training for preparedness. She sailed on 10 January 1945 and, with a fleet of battleships and cruisers, bombarded Iwo Jima on 24 January. Indiana then joined Task Force 58 at Ulithi and sortied on 10 February for the invasion of that strategic island, the next step on the island road to Japan. She supported the carriers during a raid on Tokyo on 17 February and again on 25 February, screening strikes on Iwo Jima in the interval. Indiana arrived at Ulithi for replenishment on 5 March 1945, having just supported a strike on the next target—Okinawa.

Indiana steamed out of Ulithi on 14 March for the massive Okinawa invasion and, until June 1945, steamed in support of carrier operations against Japan and Okinawa. These devastating strikes did much to aid the ground campaign and lower Japanese morale at home. During this period, she often repelled enemy suicide plane attacks as the Japanese tried desperately but vainly to stem the mounting tide of defeat. In early June, she rode out a terrific typhoon and sailed to San Pedro Bay, Philippines, on 13 June.

As a member of Task Group 38.1, Indiana operated from 1 July to 15 August, supporting air strikes against Japan and bombarding coastal targets with her big guns. The veteran battleship arrived in Tokyo Bay on 5 September and, nine days later, sailed for San Francisco, where she arrived on 29 September 1945. She was placed in reserve in commission at Bremerton on 11 September 1946. She decommissioned on 11 September 1947 and entered the Pacific Reserve Fleet. She was stricken from the Navy List on 1 June 1962 and sold for scrap. Indiana's mast is erected at the University of Indiana at Bloomington, her anchor rests at Fort Wayne, and other relics are on display in various museums and schools throughout the state.

Indiana received nine battle stars for World War II service.

BB-59 • USS MASSACHUSETTS

(BB-59: Displacement: 33,000 tons; Length: 680'10"; Beam: 108'2"; Draft: 29'3"; Speed: 27 knots; Complement: 1793; Armament: 9 16-inch, 20 5-inch, 24 40mm, 35 20mm; Class: South Dakota.)

Massachusetts (BB-59) was laid down on 20 July 1939 by the Bethlehem Steel Co., Quincy, Mass.; launched on 23 September 1941; sponsored by Mrs. Charles Francis Adams, and commissioned on 12 May 1942 at Boston, with Capt. Francis E. M. Whiting in command.

After shakedown, Massachusetts departed Casco Bay, Maine, on 24 October 1942 and, four days later, made rendezvous with the Western Naval Task Force for the invasion of North Africa, serving as flagship for Adm. H. Kent Hewitt. While steaming off Casablanca on 8 November, she came under fire from the French battleship Jean Bart's 15-inch guns. She returned fire at 0740, firing the first 16-inch shells by the U.S. against the European Axis Powers. Within a few minutes, she silenced Jean

Bart's main battery; then she turned her guns on French destroyers that had joined the attack, sinking two of them. She also shelled shore batteries and destroyed an ammunition dump. After a cease-fire had been arranged with the French, she headed for the United States on 12 November, preparing for duty in the Pacific.

Massachusetts arrived at Noumea, New Caledonia, on 4 March 1943. For the next few months, she operated in the South Pacific, protecting convoy lanes and supporting operations in the Solomons. Between 19 and 21 November, she sailed with a carrier group striking Makin, Tarawa, and Abemama in the Gilberts. On 8 December, she shelled Japanese positions on Nauru, and on 29 January 1944, she guarded carriers striking Tarawa in the Gilberts.

The navy now drove steadily across the Pacific. On 30 January, Massachusetts bombarded Kwajalein, and she covered the landings there on 1 February. With a carrier group, she struck the Japanese stronghold at Truk on 17 February. That raid inflicted heavy damage on Japanese aircraft and naval forces, and also dealt a stunning blow to enemy morale. On 21 and 22 February, Massachusetts helped fight off a heavy air attack on her task group during raids on Saipan, Tinian, and Guam. She took part in the attack on the Carolines in late March and participated in the invasion at Hollandia on 22 April, which landed 60,000 troops on the island. Retiring from Hollandia, her task force staged another attack on Truk.

Massachusetts shelled Ponape Island on 1 May, her last mission before sailing to Puget Sound for an overhaul and to reline her well-worn gun barrels. On 1 August, she left Pearl Harbor to resume operations in the Pacific war zone. Departing the Marshall Islands on 6 October, she sailed to support the landings in Leyte Gulf. In an effort to block Japanese air attacks during the Leyte conflict, she participated in a fleet strike against Okinawa on 10 October. Between 12 and 14 October, she protected forces hitting Formosa. As part of Task Group 38.3, she took part in the Battle for Leyte Gulf from 22 to 27 October, during which planes from her group sank four Japanese carriers off Cape Engano.

Stopping briefly at Ulithi, Massachusetts returned to the Philippines as part of a task force that struck Manila on 14 December while supporting the invasion of Mindoro. Massachusetts sailed into a howling typhoon on 17 December, with winds estimated at 120 knots. Three destroyers sank at the height of the typhoon's fury. Between 30 December and 23 January 1945, she sailed as part of TF 38, which struck Formosa and supported the landing at Lingayen. During that time, she turned into the South China Sea, her task force destroying shipping from Saigon to Hong Kong, concluding operations with airstrikes on Formosa and Okinawa.

From 10 February to 3 March, with the 5th Fleet, Massachusetts guarded carriers during raids on Honshu.

Her group also struck Iwo Jima by air for the invasion of that island. On 17 March, the carriers launched strikes against Kyushu while Massachusetts fired in repelling enemy attacks, splashing several planes. Seven days later, she bombarded Okinawa. She spent most of April fighting off air attacks while engaged in the operations at Okinawa, returning to the area in June when she passed through the eye of a typhoon with 100-knot winds on 5 June. She bombarded Minami Daito Jima in the Ryukyus on 10 June.

Massachusetts sailed on 1 July from Leyte Gulf to join the 3rd Fleet's final offensive against Japan. After guarding carriers launching strikes against Tokyo, she shelled Kamaishi, Honshu, on 14 July, thus hitting Japan's second-largest iron and steel center. Two weeks later, she bombarded the industrial complex at Hamamatsu, returning to blast Kamaishi on 9 August. It was here that Massachusetts fired what was probably the last 16-inch shell fired in combat in World War II.

Victory won, the fighting battleship sailed for Puget Sound for overhaul on 1 September. She left there on 28 January 1948 for operations off the California coast until leaving San Francisco for Hampton Roads, arriving on 22 April. She decommissioned on 27 March 1947 to enter the Atlantic Reserve Fleet at Norfolk and was struck from the Naval Register on 1 June 1962.

"Big Mamie," as she was affectionately known, was saved from the scrap pile when she was transferred to the Massachusetts Memorial Committee on 8 June 1965. She was enshrined at Fall River, Mass., on 14 August 1965, as the Bay State's memorial to those who gave their lives in World War II.

Massachusetts received 11 battle stars for World War II service.

BB-60 • USS ALABAMA

(BB 60: dp. 35, 000, 1. 680', b. 108'2", dr. 36'2", s. 27.5 k., cpl. 1, 793; a. 9 16", 20 5"; 24 40mm., Z 20mm.; cl. South Dakota)

The third ship named Alabama (BB 60) was laid down on 1 February 1940 by the Norfolk (Va.) Navy Yard was launched on 16 February 1942, sponsored by Mrs. Lister Hill, wife of the senior senator from Alabama, and commissioned on 16 August 1942, Capt. George B. Wilson in command.

After fitting out, Alabama commenced her shakedown cruise in Chesapeake Bay on Armistice Day (11 November) 1942. As 1943 began, the new battleship headed north to conduct operational training out of Casco Bay, Maine. She returned to Chesapeake Bay on 11 January 1943 to carry out the last week of shakedown training. Following a period of availability and logistics support at Norfolk, Alabama, was assigned to Task Group (TG) 22.2 and returned to Casco Bay for tactical maneuvers on 13 February 1943.

With the movement of substantial British strength toward the Mediterranean theater to prepare for the invasion of Sicily, the Royal Navy lacked the heavy ships necessary to cover the northern convoy routes. The British appeal for help on those lines soon led to the temporary assignment of Alabama and South Dakota (BB 57) to the Home Fleet.

On 2 April 1943, Alabama-as part of Task Force (TF) 22— sailed for the Orkney Islands with her sister ship and a screen of five destroyers. Proceeding via Little Placentia Sound Argentia, Newfoundland, the battleship reached Scapa Flow on 19 May 1943, reporting for duty with TF 61 and becoming a unit of the British Home Fleet. She soon embarked on intensive operational training to coordinate joint operations.

Early in June, Alabama and her sister ship, along with British Home Fleet units, covered the reinforcement of the garrison on the island of Spitzbergen, which lay on the northern flank of the convoy route to Russia, in an operation that took the ship across the Arctic Circle. Soon after returning to Scapa Flow, Admiral Harold R. Stark, Commander of the United States Naval Forces, Europe, inspected her.

Shortly thereafter, in July, Alabama participated in Operation "Governor," a diversion aimed toward southern Norway to draw German attention away from the real Allied thrust toward Sicily. It had also been devised to attempt to lure out the German battleship Tirpitz, the sister ship of the famed but short-lived Bismarck, but the Germans did not rise to the challenge, and the enemy battleship remained in her Norwegian lair.

Alabama was detached from the British Home Fleet on 1 August 1943 and, in company with South Dakota and screening destroyers, sailed for Norfolk, arriving there on August 9. For the next ten days, Alabama underwent a period of overhaul and repairs. This work was completed, and the battleship departed Norfolk on 20 August 1943 for the Pacific. Transiting the Panama Canal five days later, she dropped anchor in Havannah Harbor, at Efate in the New Hebrides, on 14 September.

Following a month and a half of exercises and training with fast carrier task groups, the battleship moved to Fiji on 7 November. Alabama sailed on 11 November to take part in Operation "Galvanic," the assault on the Japanese-held Gilbert Islands. She screened the fast carriers as they launched attacks on Jaluit and Mille atoms, Marshall Islands, to neutralize Japanese airfields located there. Alabama supported landings on Tarawa on 20 November and later took part in the securing of Betio and Makin. On the night of 26 November, Alabama twice opened fire to drive off enemy aircraft that approached her formation.

On 8 December 1943, Alabama, along with five other fast battleships, carried out the first Pacific gunfire strike conducted by that type of warship. Alabama's guns hurled 535 rounds into enemy strongpoints as she and her sister ships bombarded Nauru Island, an enemy phosphate-producing center, causing severe damage to shore installations there. She also took the destroyer Boyd (DD 544) alongside after that ship had received a direct hit from a Japanese shore battery on Nauru and brought three injured men on board for treatment.

She then escorted the carriers Bunker Hill (CV-17) and Monterey (CVL-26) back to Efate, arriving on 12 December. Alabama departed the New Hebrides for Pearl Harbor on 5 January 1944, arrived on the 12th, and underwent a brief drydocking at the Pearl Harbor Navy Yard. After the replacement of her port outboard propeller and routine maintenance, Alabama was again underway to return to action in the Pacific.

Alabama reached Funafuti, Ellice Islands, on 21 January 1944 and rejoined the Fleet there. Assigned to Task Group (TG) 58.2, which was formed around Essex (CV-9), Alabama, left the Ellice Islands on 25 January to help carry out Operation "Flintlock," the invasion of the Marshall Islands. Alabama, along with sister ship South Dakota and the fast battleship North Carolina (BB-55), bombarded Roi on 29 January and Namur on 30 January. She hurled 330 rounds of 16-inch and 1 562 of 5-inch toward Japanese targets, destroying planes, airfield facilities, blockhouses, buildings, and gun emplacements. Over the following days of the campaign, Alabama patrolled the area north of Kwajalein Atoll. On 12 February 1944, Alabama sortied with the Bunker Hill task group to launch attacks on Japanese installations, aircraft, and shipping at Truk. Those raids, launched on February 16 and 17, caused heavy damage to enemy shipping concentrated at that island base.

Leaving Truk, Alabama began steaming toward the Marianas to assist in strikes on Tinian, Saipan, and Guam. During this action, while repelling enemy air attacks on 21 February 1944, 5-inch Mount No. 9 accidentally fired into Mount No. 5. Five men died, and 11 were wounded in the mishap.

After the strikes were completed on 22 February, Alabama conducted a sweep looking for crippled enemy ships southeast of Saipan and eventually returned to Majuro on 26 February 1944. There, she served temporarily as flagship for Vice Admiral Marc A. Mitscher, Commander, TF 58, from 3 to 8 March.

Alabama's next mission was to screen the fast carriers as they hurled air strikes against Japanese positions on Palau, Yap Ulithi, and Woleai, Caroline Islands. She steamed from Majuro on 22 March 1944 with TF 58 in the screen of Yorktown (CV-10). On the night of 29 March, about six enemy planes approached TG 58.3, in which Alabama was operating, and four broke off to attack ships in the vicinity of the battleship. Alabama downed one unassisted and helped in the destruction of another.

On 30 March, planes from TF 58 began bombing Japanese airfields, shipping, fleet servicing facilities, and other installations on the islands of Palau, Yap, Ulithi, and Woleai. During that day, Alabama again provided antiaircraft fire whenever enemy planes appeared. At 2045 on the 30th, a single plane approached TG 58.3, but Alabama and other ships drove it off before it could cause any damage.

The battleship returned briefly to Majuro before she sailed on 13 April with TF 58, this time in the screen of Enterprise (CV -). In the next three weeks, TF 58 hit enemy targets on Hollandia Wake, Sawar, and Sarmi along the New Guinea coast, covered Army landings at Aitape, Tanahmerah Bay, and Humboldt Bay, and conducted further strikes on Truk.

As part of the preliminaries to the invasion of the Marianas, Alabama, in company with five other fast battleships, shelled the large island of Ponape in the Carolines, the site of a Japanese airfield and seaplane base. As Alabama's Caut. Fred T Kirtland subsequently noted the bombardment, of 70 minutes duration, was conducted in a

"leisurely manner. " Alabama then returned to Majuro on 4 May 1944 to prepare for the invasion of the Marianas.

After a month spent in exercises and refitting, Alabama again got underway with TF 58 to participate in Operation "Forager. " On 12 June, Alabama screened the carriers striking Saipan. On 13 June, Alabama took part in a six-hour pre-invasion bombardment of the west coast of Saipan to soften the defenses and cover the initial minesweeping operations. Her spotting planes reported that her salvoes had caused great destruction and fires in Garapan town. Though the shelling appeared successful, it proved to be a failure due to the lack of specialized training and experience required for a successful shore bombardment. Strikes continued as troops invaded Saipan on 15 June.

On 19 June, during the Battle of the Philippine Sea, Alabama operated with TG 58.7, providing antiaircraft support for the fast carriers against attacking Japanese aircraft. The ships of TF 58 claimed 27 enemy planes downed during the course of the action, which later came to be known as the "Marianas Turkey Shoot."

In the first raid that approached Alabama's formation, only two planes managed to penetrate to attack her sistership, South Dakota, scoring one bomb hit that caused minor damage. An hour later, a second wave, composed largely of torpedo bombers, bore in, but Alabama's barrage discouraged two planes from attacking South Dakota. The intense concentration paid to the incoming torpedo planes left one dive bomber nearly undetected, and it managed to drop its load near Alabama; the two small bombs were near-misses and caused no damage.

American submarines sank two Japanese carriers, and Navy pilots claimed a third carrier. American pilots and antiaircraft gunners had seriously depleted Japanese naval air power. Out of the 430 planes with which the enemy had commenced the Battle of the Philippine Sea, only 35 remained operational afterward.

Alabama continued patrolling areas around the Marianas to protect the American landing forces on Saipan, screening the fast carriers as they struck enemy shipping, aircraft, and shore installations on Guam, Tinian, Rota, and Saipan. She then retired to the Marshalls for upkeep.

Commander Battleship Division 9—left Eniwetok on 14 July 1944, sailing with the task group formed around Bunker Hill. She screened the fast carriers as they conducted pre-invasion attacks and supported the landings on the island of Guam on 21 July. She returned briefly to Eniwetok on 11 August. On 30 August, she got underway in the screen of Essex to carry out Operation "Stalemate II, the seizure of Palau, Ulithi, and Yap. On September 6 through 8, the forces launched strikes on the Carolines.

Alabama departed the Carolines to sail to the Philippines and provided cover for the carriers striking the islands of Cebu Leyte, Bohol, and Negros from 12 to 14 September. The carriers launched strikes on shipping and installations in the Manila Bay area on 21 and 22 September and in the central Philippines area on 24 September. Alabama retired briefly to Salpan on 28 September, then proceeded to Ulithi on 1 October 1944.

On 6 October 1944, Alabama sailed with TF 38 to support the liberation of the Philippines. Again operating as part of a fast carrier task group, Alabama protected the flattops while they launched strikes on Japanese facilities at Okinawa, in the Pescadores, and in Formosa.

Detached from the Formosa area on 14 October to sail toward Luzon, the fast battleship again used her antiaircraft batteries in support of the carriers as enemy aircraft attempted to attack the formation. Alabama's gunners claimed three enemy aircraft shot down and a fourth damaged. By 15 October, Alabama was supporting landing operations on Leyte. She then screened the carriers as they conducted air strikes on Cebu, Negros, Panay, northern Mindanao, and Leyte on 21 October 1944.

Alabama, as part of the Enterprise screen, supported air operations against the Japanese Southern Force in the area off Surigao Straut, then moved north to strike the powerful Japanese Central Force heading for San Bernardino Strait. After receiving reports of a third Japanese force, the battleship served on the screen of the fast carrier task force as it sped to Cape Engano. On 24 October, although American air strikes destroyed four Japanese carriers in the Battle of Cape Engano, the Japanese Central Force under Admiral Kurita had transited San Bernardino Strait and emerged off the coast of Samar, where it fell upon a task group of American escort carriers and their destroyer and destroyer escort screen. Alabama reversed her course and headed for Samar to assist the greatly outnumbered American forces, but the Japanese had retreated by the time she

reached the scene. She then joined the protective screen for the Essex task group to hit enemy forces in the central Philippines before retiring to Ulithi on 30 October 1944 for replenishment.

Underway again on 3 November 1944, Alabama screened the Fast carriers carried out sustained strikes against Japanese airfields and installations on Luzon to prepare for a landing on Mindoro Island. She spent the next few weeks engaged in operating against the Visayas and Luzon before retiring to Ulithi on 24 November.

The first half of December 1944 found Alabama engaged in various training exercises and maintenance routines. She left Ulithi on 10 December and reached the launching point for air strikes on Luzon on 14 December as the fast carrier task forces launched aircraft to carry out preliminary strikes on airfields on Luzon that could threaten the landings slated to take place on Mindoro. From 14 to 16 December, a veritable umbrella of carrier aircraft covered the Luzon fields, preventing any enemy planes from getting airborne to challenge the Mindoro-bound convoys. Having completed her mission, she left the area to refuel on 17 December, but as she reached the fueling rendezvous, she began encountering heavy weather. By daybreak on the 18th, rough seas and harrowing conditions rendered fueling at sea impossible; 50-knot winds caused ships to roll heavily. Alabama experienced rolls of 30 degrees, had both her Vought ''Kingfisher''floatplanes so badly damaged that they were of no further value, and received minor damage to her structure. At one point in the typhoon, Alabama recorded wind gusts up to 83 knots. Three destroyers, Hull (DD-350), Monaghan (DD 354), and Spence (DW-512), were lost to the typhoon. By 19 December, the storm had run its course, and Alabama arrived back at Ulithi on 24 December. After pausing there briefly, Alabama continued on to Puget Sound Naval Shipyard for overhaul.

The battleship entered drydock on 18 January 1945 and remained there until 25 February. Work continued until 17 March, when Alabama got underway for standardization trials and refresher training along the southern California coast. She got underway for Pearl Harbor on April 4 and arrived there on April 10.

And held a week of training exercises. She then continued on to Ulithi and moored there on April 28, 1945.

Alabama departed Ulithi with TF 58 on 9 May 1945, bound for the Ryukyus, to support forces that had landed on Okinawa on 1 April 1945 and to protect the fast carriers as they launched air strikes on installations in the Ryukyus and on Kyushu. On 14 May, several Japanese planes penetrated the combat air patrol to get at the carriers; one crashed Vice Admiral Mitscher's flagship. Alabama's guns splashed two and assisted in splashing two more.

Subsequently, Alabama rode out a typhoon on June 4 and 5, suffering only superficial damage, while the nearby heavy cruiser Pittsburgh (CA-70) lost her bow. Alabama subsequently bombarded the Japanese island of Minami Daito Shima, with other fast battleships, on 10 June 1945 and then headed for Leyte Gulf later in June to prepare to strike at the heart of Japan with the 3d Fleet.

On 1 July 1945, Alabama and other 3d Fleet units got underway for the Japanese home islands. Throughout the month of July 1945, Alabama carried out strikes on targets in industrial areas of Tokyo and other points on Honshu, Hokkaido, and Kyushu; on the night of 17 and 18 July, Alabama and other fast battleships in the task group carried out the first night bombardment of six major industrial plants in the Hitachi-Mito area of Honshu, about eight miles northeast of Tokyo. On board Alabama to observe the operation was retired Rear Admiral Richard E. Byrd, the famed polar explorer.

On 9 August, Alabama transferred a medical party to the destroyer Ault (DD-98) for further transfer to the destroyer Borze (DD-704). The latter had been kamikazied on that date and required prompt medical aid at her distant picket station.

At the end of the war, Alabama was still at sea, operating off the southern coast of Honshu. On 15 August 1945, she received word of the Japanese capitulation. During the initial occupation of the Yokosuka-Tokyo area, Alabama transferred detachments of marines and bluejackets for temporary duty ashore; her bluejackets were among the first from the Fleet to land. She also served on the screen of the carriers as they conducted reconnaissance flights to locate prisoner-of-war camps.

Alabama entered Tokyo Bay on 5 September to receive men who had served with the occupation forces and then departed Japanese waters on 20 September. At Okinawa, she embarked 700 sailors—principally members of Navy construction battalions (or "Seabees"—for her part in the "Magic Carpet" operations. She reached San Francisco at midday on 15 October, and on Navy Day (27 October 1945), she hosted 9, 000 visitors. She then shifted to San Pedro, Calif., on 29 October. Alabama remained at San Pedro through 27 February 1946, when she left for the Puget Sound Naval Shipyard for an inactivation overhaul. Alabama was decommissioned on 9 January 1947 at the Naval Station, Seattle, and was assigned to the Bremerton Group, United States Pacific Reserve Fleet. She remained there until struck from the Naval Vessel Register on 1 June 1962.

Citizens of the state of Alabama had formed the "USS Alabama Battleship Commission" to raise funds for the preservation of Alabama as a memorial to the men and women who served in World War II. The ship was awarded to that state on 16 June 1964 and was formally turned over on 7 July 1964 in ceremonies at Seattle. Alabama was then towed to her permanent berth at Mobile, Ala., arriving in Mobile Bay on 14 September 1964.

Alabama received nine battle stars for her World War II service.

BB-61 • USS IOWA

(BB-61: dp. 45,000 t.; l. 887'3"; b. 108'2"; dr. 37'9"; s. 33 k.; cpl. 2,800; a. 9 16", 20 5"; cl. Iowa)

The third ship named Iowa (BB-61) was laid down at the New York Navy Yard on 27 June 1940, launched on 27 August 1942, sponsored by Mrs. Henry A. Wallace, wife of Vice President Wallace, and commissioned on 22 February 1943, with Captain John L. McCrea in command.

On 24 February, Iowa put to sea for shakedown in Chesapeake Bay and along the Atlantic coast. She got underway on 27 August for Argentia, Newfoundland to neutralize the threat of the German Battleship Tirpitz, which was reportedly operating in Norwegian waters.

In the fall, Iowa carried President Franklin D. Roosevelt to Casablanca, French Morocco, on the first leg of his journey to the Tehran Conference in November. After the conference, she returned the President to the United States.

As Flagship of Battleship Division 7, Iowa departed the United States on 2 January 1944 for the Pacific Theatre and her combat debut in the campaign for the Marshall Islands. From 29 January to 3 February, she supported carrier air strikes made by Rear Admiral Frederick C. Sherman's task group against Kwajalein and Eniwetok Atolls in the Marshall Islands. Her next assignment was to support air strikes against the Japanese Naval base at Truk, Caroline Islands. Iowa, in company with other ships, was detached from the support group on 16 February 1944 to conduct an anti-shipping sweep around Truk to destroy enemy naval vessels escaping to the north. On 21 February, she was underway with the Fast Carrier Task Force 58 while it conducted the first strikes against Saipan, Tinian, Rota, and Guam in the Marianas.

On 18 March, Iowa, flying the flag of Vice Admiral Willis A. Lee, Commander Battleships, Pacific, joined in the bombardment of Mili Atoll in the Marshall Islands. Although struck by two Japanese 4.7" projectiles during

the action, Iowa suffered negligible damage. She then rejoined Task Force 58 on 30 March and supported air strikes against the Palau Islands and Woleai in the Carolines, which continued for several days.

From 22 to 28 April 1944, Iowa supported air raids on Hollandia, Aitape, and Wakde Islands to support Army forces on Aitape, Tanahmerah Bay, and Humboldt Bay in New Guinea. She then joined the Task Force's second strike on Truk, 29-30 April, and bombarded Japanese facilities on Ponape in the Carolines on 1 May.

In the opening phases of the Marianas campaign, Iowa protected the carriers during air strikes on the islands of Saipan, Tinian, Guam, Rota, and Pagan on 12 June. Iowa was then detached to bombard enemy installations on Saipan and Tinian on 13-14 June. On 19 June, in an engagement known as the Battle of the Philippine Sea, Iowa, as part of the battle line of Fast Carrier Task Force 58, helped repel four massive air raids launched by the Japanese Middle Fleet, resulting in the almost complete destruction of Japanese carrier-based aircraft. Iowa then joined in the pursuit of the fleeing enemy Fleet, shooting down one torpedo plane and assisting in splashing another.

Throughout July, Iowa remained off the Marianas supporting air strikes on the Palaus and landings on Guam. After a month's rest, Iowa sortied from Eniwetok as part of the 3rd Fleet and helped support the landings on Peleliu on 17 September. She then protected the carriers during air strikes against the Central Philippines to neutralize enemy air power for the long-awaited invasion of the Philippines. On 10 October, Iowa arrived off Okinawa for a series of air strikes on the Ryukyus and Formosa. She then supported air strikes against Luzon on

15 October and continued this vital duty during General MacArthur's landing on Leyte on 20 October.

In a last-ditch attempt to halt the United States campaign to recapture the Philippines, the Japanese Navy struck back with a three-pronged attack aimed at the destruction of American amphibious forces in Leyte Gulf. Iowa accompanied TF 38 during attacks against the Japanese Central Force as it steamed through the Sibuyan Sea toward San Bernardino Strait. The reported results of these attacks and the apparent retreat of the Japanese Central Force led Admiral Halsey to believe that this force had been ruined as an effective fighting group. Iowa, with Task Force 38, steamed after the Japanese Northern Force off Cape Engaño, Luzon. On 25 October 1944, when the ships of the Northern Force were almost within range of Iowa's guns, word arrived that the Japanese Central Force was attacking a group of American escort carriers off Samar. This threat to the American beachheads forced her to reverse course and steam to support the vulnerable "baby carriers." However, the valiant fight put up by the escort carriers and their screen had already caused the Japanese to retreat, and Iowa was denied a surface action. Following the Battle for Leyte Gulf, Iowa remained in the waters off the Philippines, screening carriers during strikes against Luzon and Formosa. She sailed for the West Coast late in December 1944.

Iowa arrived in San Francisco on 15 January 1945 for overhaul. She sailed on 19 March 1945 for Okinawa, arriving on 15 April 1945. Commencing 24 April 1945, Iowa supported carrier operations which assured American troops vital air superiority during their struggle for that bitterly contested island. She then supported air

strikes off southern Kyushu from 25 May to 13 June 1945. Iowa participated in strikes on the Japanese homeland on 14-15 July and bombarded Muroran, Hokkaido, destroying steel mills and other targets. The city of Hitachi on Honshu was given the same treatment on the night of 17-18 July 1945. Iowa continued to support fast carrier strikes until the cessation of hostilities on 15 August 1945.

Iowa entered Tokyo Bay with the occupation forces on 29 August 1945. After serving as Admiral William F. Halsey's flagship for the surrender ceremony on 2 September 1945, Iowa departed Tokyo Bay on 20 September 1945 for the United States.

Arriving in Seattle, Wash., on 15 October 1945, Iowa returned to Japanese waters in January 1946 and became the flagship of the 5th Fleet. She continued this role until she sailed for the United States on 25 March 1946. From that time until September 1948, Iowa operated from West Coast ports, engaging in Naval Reserve and at-sea training, as well as drills and maneuvers with the Fleet. Iowa decommissioned on 24 March 1949. After Communist aggression in Korea necessitated an expansion of the active fleet, Iowa was recommissioned on 25 August 1951, with Captain William R. Smedberg III in command. She operated off the West Coast until March 1952, when she sailed for the Far East. On 1 April 1952, Iowa became the flagship of Vice Admiral Robert T. Briscoe, Commander, 7th Fleet, and departed Yokosuka, Japan, to support United Nations Forces in Korea. From 8 April to 16 October 1952, Iowa was involved in combat operations off the East Coast of Korea. Her primary mission was to aid ground troops by bombarding enemy targets at Songjin, Hungnam, and Kojo, North Korea. During this

time, Admiral Briscoe was relieved as Commander, 7th Fleet, by Vice Admiral J. J. Clark, who continued to use Iowa as his flagship until 17 October 1952. Iowa departed Yokosuka, Japan, on 19 October 1952 for overhaul at Norfolk and training operations in the Caribbean Sea.

Iowa embarked midshipmen for at-sea training to Northern Europe in July 1953, and immediately afterward took part in Operation "Mariner," a major NATO exercise, serving as the flagship of Vice Admiral E. T. Woolfidge, commanding the 2nd Fleet. Upon completion of this exercise, until the fall of 1954, Iowa operated in the Virginia Capes area. In September 1954, she became the flagship of Rear Admiral R. E. Libby, Commander, Battleship-Cruiser Force, U.S. Atlantic Fleet.

From January to April 1955, Iowa made an extended cruise to the Mediterranean as the first battleship regularly assigned to the Commander, 6th Fleet. Iowa departed on a midshipman training cruise on 1 June 1955 and upon her return, she entered Norfolk for a 4-month overhaul. Following refit, Iowa continued intermittent training cruises and operational exercises until 4 January 1957, when she departed Norfolk for duty with the 6th Fleet in the Mediterranean. Upon completion of this deployment, Iowa embarked midshipmen for a South American training cruise and joined in the International Naval Review off Hampton Roads, Va., on 13 June 1957.

On 3 September 1957, Iowa sailed for Scotland for NATO Operation "Strikeback." She returned to Norfolk on 28 September 1957 and departed Hampton Roads for the Philadelphia Naval Shipyard on 22 October 1957. She decommissioned on 24 February 1958 and entered the Atlantic Reserve Fleet at Philadelphia.

In the 1980s, reflecting a shift in U.S. naval policy under President Reagan's administration, the USS Iowa was

reactivated and underwent substantial modernization. This included the addition of contemporary weapon systems, such as Tomahawk missiles and Phalanx close-in weapon systems, which drastically enhanced her combat capabilities.

During her reactivated service, the USS Iowa was involved in several notable incidents. In 1984, she was deployed to the Lebanese Civil War, where she conducted bombardments against Druze and Syrian positions in the Beqaa Valley east of Beirut. This operation was a significant example of naval gunfire support during the post-World War II era.

A tragic event marked the USS Iowa's history on April 19, 1989, when an explosion in one of her 16-inch gun turrets killed 47 crew members. This incident brought about intensive investigations and led to changes in ship safety and munitions handling procedures.

The USS Iowa was decommissioned on October 26, 1990, as the Cold War drew to a close and the strategic need for such large battleships decreased. This marked the end of an era for the battleships that had been central to naval warfare in the first half of the 20th century.

In her post-service life, the USS Iowa found a new role as a museum ship. She was donated to the Pacific Battleship Center and is now permanently moored at the Port of Los Angeles in San Pedro, California. As a museum, the USS Iowa serves as a living exhibit, preserving the history of American naval power and the evolution of naval warfare.

BB-62 • USS NEW JERSEY

(BB-62: displacement 45,000; l. 887'7", beam 108'1", draft 28'11", speed 33 k.; complement 1,921; armament 9 16", 20 5"; class Iowa)

The second ship named New Jersey (BB-62) was launched on 7 December 1942 by the Philadelphia Naval Shipyard; sponsored by Mrs. Charles Edison, wife of Governor Edison of New Jersey, former Secretary of the Navy, and commissioned at Philadelphia on 23 May 1943, Captain Carl F. Holden in command.

New Jersey completed fitting out and trained her initial crew in the Western Atlantic and Caribbean. On 7 January 1944, she passed through the Panama Canal warbound for Funafuti, Ellice Islands. She reported there on 22 January for duty with the Fifth Fleet, and three days later rendezvoused with Task Group 58.2 for the assault on the Marshall Islands. New Jersey screened the carriers from enemy attack as their aircraft flew strikes against Kwajalein and Eniwetok from 29 January to 2 February, softening up the latter for its invasion and supporting the troops who landed on 31 January.

New Jersey began her distinguished career as a flagship on 4 February in Majuro Lagoon when Admiral Raymond A. Spruance, commanding the Fifth Fleet, broke his flag from her mast. Her first action as a flagship was a bold two-day surface and air strike by her task force against the supposedly impregnable Japanese fleet base on Truk in the Carolines. This blow was coordinated with the assault on Kwajalein and effectively interdicted Japanese naval retaliation to the conquest of the Marshalls. On 17 and 18 February, the task force accounted for two Japanese light cruisers, four destroyers, three auxiliary cruisers, two submarine tenders, two submarine chasers, an armed trawler, a plane ferry, and 23 other auxiliaries, not including small craft. New Jersey destroyed a trawler and, with other ships, sank the destroyer Maikaze, as well as firing on an enemy plane that attacked her formation. The task force returned to the Marshalls on 19 February.

Between 17 March and 10 April, New Jersey first sailed with Rear Admiral Marc A. Mitscher's flagship Lexington (CV-16) for an air and surface bombardment of Mili, then rejoined Task Group 58.2 for a strike against shipping in the Palau Islands and bombarded Woleai. Upon his return to Majuro, Admiral Spruance transferred his flag to Indianapolis (CA-35).

New Jersey's next war cruise, from 13 April to 4 May, began and ended at Majuro. She screened the carrier striking force which provided air support to the invasion of Aitape, Tanahmerah Bay, and Humboldt Bay, New Guinea, on 22 April, then bombed shipping and shore installations at Truk on 29-30 April. New Jersey and her formation splashed two enemy torpedo bombers at Truk. Her sixteen-inch salvos pounded Ponape on 1 May, destroying fuel tanks, badly damaging the airfield, and demolishing a headquarters building.

After rehearsing in the Marshalls for the invasion of the Marianas, New Jersey put to sea on 6 June in the screening and bombardment group of Admiral Mitscher's Task Force. On the second day of pre-invasion air strikes, 12 June, New Jersey downed an enemy torpedo bomber, and during the next two days, her heavy guns battered Saipan and Tinian, throwing steel against the beaches the Marines would charge on 15 June.

The Japanese response to the Marianas operation was an order to its Mobile Fleet: it must attack and annihilate the American invasion force. Shadowing American submarines tracked the Japanese fleet into the Philippine Sea as Admiral Spruance joined his task force with Admiral Mitscher's to meet the enemy. New Jersey took station in the protective screen around the carriers on 19 June as American and Japanese pilots dueled in the Battle of the Philippine Sea. That day and the next were to pronounce the doom of Japanese naval aviation; in this "Marianas Turkey Shoot," the Japanese lost some 400 planes. This loss of trained pilots and aircraft was equaled in disaster by the sinking of three Japanese carriers by submarines and aircraft, and the damaging of two carriers and a battleship. The anti-aircraft fire of New Jersey and the other screening ships proved virtually impenetrable. Only two American ships were damaged, and those but slightly. In this overwhelming victory, only 17 American planes were lost to combat.

New Jersey's final contribution to the conquest of the Marianas was in strikes on Guam and the Palaus, from which she sailed for Pearl Harbor, arriving on 9 August. Here she broke the flag of Admiral William F. Halsey, Jr., on 24 August, becoming the flagship of the Third Fleet. For the eight months after she sailed from Pearl Harbor on 30 August, New Jersey was based at Ulithi. In this climactic span of the Pacific War, fast carrier task forces ranged the waters off the Philippines, Okinawa, and Formosa, striking again and again at airfields, shipping, shore bases, and invasion beaches. New Jersey offered the essential protection required by these forces, always ready to repel enemy air or surface attacks.

In September, the targets were in the Visayas and the southern Philippines, then Manila and Cavite, Panay, Negros, Leyte, and Cebu. Early in October, raids to destroy enemy air power based on Okinawa and Formosa were begun in preparation for the Leyte landings on 20 October.

This invasion brought on the desperate, almost suicidal, last great sortie of the Imperial Japanese Navy. Its plan for the Battle for Leyte Gulf included a feint by a northern force of planeless heavy attack carriers to draw away the battleships, cruisers, and fast carriers with which Admiral Halsey was protecting the landings. This was to allow the Japanese Center Force to enter the gulf through San Bernardino Strait. At the opening of the battle, planes from the carriers guarded by New Jersey struck hard at both the Japanese Southern and Center Forces, sinking a battleship on 23 October. The next day, Halsey shaped his course north after the decoy force had been spotted. Planes from his carriers sank four of the Japanese carriers, as well as a destroyer and a cruiser, while New Jersey steamed south at flank speed to meet the newly developed threat of the Center Force. It had been turned back in a stunning defeat when she arrived.

New Jersey rejoined her fast carriers near San Bernardino on 27 October for strikes on central and southern Luzon. Two days later, the force was under a suicide attack. In a melee of anti-aircraft fire from the ships and combat air patrol, New Jersey shot down a plane whose pilot maneuvered it into Intrepid's (CV-11) port gun galleries, while machine gun fire from Intrepid wounded three of New Jersey's men. During a similar action on 25 November, three Japanese planes were splashed by the combined fire of the force, part of one crashing onto Hancock's (CV-19) flight deck. Intrepid was again attacked, shot down one would-be suicide, but was crashed by another despite hits scored on the attacker by New Jersey gunners. New Jersey shot down a plane diving on Cabot (CVL-28) and hit another which smashed into Cabot's port bow.

In December, New Jersey sailed with the Lexington task group for air attacks on Luzon from 14 to 16 December, then found herself in the furious typhoon which sank three destroyers. Skillful seamanship brought her through undamaged. She returned to Ulithi on Christmas Eve to be met by Fleet Admiral Chester W. Nimitz.

New Jersey ranged far and wide from 30 December to 25 January 1945 on her last cruise as Admiral Halsey's flagship. She guarded the carriers in their strikes on Formosa, Okinawa, and Luzon, on the coast of Indochina, Hong Kong, Swatow, and Amoy, and again on Formosa and Okinawa. At Ulithi on 27 January, Admiral Halsey lowered his flag in New Jersey, but it was replaced two days later by that of Rear Admiral Oscar Badger commanding Battleship Division Seven.

In support of the assault on Iwo Jima, New Jersey screened the Essex (CV-9) group in air attacks on the island from 19 to 21 February and provided the same crucial service for the first major carrier raid on Tokyo on 25 February, a raid aimed specifically at aircraft production. During the next two days, Okinawa was attacked from the air by the same striking force.

New Jersey was directly engaged in the conquest of Okinawa from 14 March until 16 April. As the carriers prepared for the invasion with strikes there and on Honshu, New Jersey fought off air raids, used her seaplanes to rescue downed pilots, defended the carriers from suicide planes, shooting down at least three and assisting in the destruction of others. On 24 March, she again carried out the vital battleship role of heavy bombardment, preparing the invasion beaches for the assault a week later.

During the final months of the war, New Jersey underwent an overhaul at Puget Sound Naval Shipyard. Departing on July 4, she headed to San Pedro, Pearl Harbor, Eniwetok, and ultimately Guam. On August 14, she again became the flagship of the Fifth Fleet under Admiral Spruance, anchoring in Manila and Okinawa before reaching Tokyo Bay on September 1. There, she served as the flagship for successive commanders of Naval Forces in Japanese waters until Iowa (BB-61) relieved her on January 28, 1946. Carrying nearly a thousand homeward-bound troops, she arrived in San Francisco on February 10.

After operations along the West Coast and a routine overhaul at Puget Sound, New Jersey crossed the Atlantic to celebrate her fourth birthday in Bayonne, New Jersey, on May 23, 1947. The event was attended by Governor Alfred E. Driscoll, former Governor Walter E. Edge, and other dignitaries.

From June 7 to August 26, New Jersey joined the first training squadron to cruise Northern European waters since World War II began. Over two thousand Naval Academy and NROTC midshipmen gained seagoing experience under Admiral Richard L. Connoly, Commander of Naval Forces Eastern Atlantic and Mediterranean. Official receptions were held in Oslo, where King Haakon VII of Norway inspected the crew on July 2, and in Portsmouth, England. The training fleet headed westward on July 18 for exercises in the Caribbean and Western Atlantic.

Serving as the flagship for Rear Admiral Heber H. McLean, Commander of Battleship Division One, in New York from September 12 to October 18, New Jersey was subsequently inactivated at the New York Naval Shipyard. She was decommissioned at Bayonne on June 30, 1948, and joined the New York Group of the Atlantic Reserve Fleet.

Recommissioned at Bayonne on November 21, 1950, with Captain David M. Tyree in command, New Jersey prepared for the European War in the Caribbean. She departed from Norfolk on April 16, 1951, and arrived off the east coast of Korea on May 17. Vice Admiral Harold M. Martin, commanding the Seventh Fleet, hoisted his flag on New Jersey for six months.

New Jersey's Korean War service included extensive shore bombardments, supporting United Nations troops, and acting as mobile artillery. She sustained combat casualties, including one fatality, when hit by a shore battery on her number one turret. The battleship engaged in multiple operations, destroying enemy supply routes, troop positions, and providing vital support in various battles.

After serving in Korean waters, New Jersey returned to the United States for overhauls and training. She

continued to play an active role in Naval operations, including serving as flagship for various admirals and participating in training cruises. She was decommissioned and placed in reserve at Bayonne on August 21, 1957.

New Jersey was recommissioned on April 6, 1968, at Philadelphia Naval Shipyard, with Captain J. Edward Snyder in command. Tailored for heavy bombardment, she saw action in Vietnam, demonstrating the power of her 16-inch guns. After a series of deployments and operations, New Jersey was decommissioned on December 17, 1969, and joined the inactive fleet.

On December 28, 1982, New Jersey was recommissioned at Long Beach, California, marking a return of the world's last battleships. Her later service included operations in the Bekaa Valley, as part of a Pacific Fleet battleship group, and participation in various exercises and celebrations. She was involved in the Persian Gulf during the late 1980s and was finally decommissioned following her last operational cruise in 1990.

After her final decommissioning, the USS New Jersey (BB-62) found a new life as a museum ship. She was donated to the Home Port Alliance of Camden, New Jersey, for use as a museum. On October 15, 2001, she arrived at her final resting place on the Delaware River in Camden, New Jersey. Since then, the USS New Jersey has been open to the public as a museum and memorial, showcasing her rich history and the significant role she played in American naval warfare.

As a museum, the USS New Jersey offers visitors a chance to explore numerous aspects of the ship, including the main deck, various gun turrets, and living quarters, providing a glimpse into the life of sailors during her years of service. The museum also hosts educational programs, overnight encampments, and special events, preserving the legacy of this iconic battleship and those who served aboard her.

New Jersey earned the Navy Unit Commendation for her service in Vietnam, along with numerous battle stars for her involvement in World War II, the Korean conflict, and Vietnam.

BB-63 • USS MISSOURI

(BB-63: displacement 45,000; length 887'3"; beam 108'2"; draft 28'11"; speed 33 knots; complement 1,921; armament 9 16-inch guns, 20 5-inch guns; class Iowa)

The fourth ship named Missouri (BB-63), the last battleship completed by the United States, was laid down on January 6, 1941, by the New York Naval Shipyard. It was launched on January 29, 1944, sponsored by Miss Margaret Truman, daughter of then Senator from Missouri Harry S. Truman, later President; and commissioned on June 11, 1944, with Captain William M. Callaghan in command.

After trials off New York and shakedown and battle practice in Chesapeake Bay, Missouri departed Norfolk on November 11, transited the Panama Canal on November 18, and steamed to San Francisco for final fitting out as fleet flagship. She left San Francisco Bay on December 14 and arrived at Ulithi, West Caroline Islands, on January 13, 1945. There, she served as the temporary headquarters ship for Vice Admiral Marc A. Mitscher. The battleship set sail on

January 27 to join the screen of the Lexington carrier task group of Mitscher's Task Force 58, and on February 16, her aircraft carriers launched the first airstrikes against Japan since the Doolittle raid launched from the carrier Hornet in April 1942.

Missouri then accompanied the carriers to Iwo Jima, where her guns provided direct and continuous support to the invasion landings beginning on February 19. After TF 58 returned to Ulithi on March 5, Missouri was assigned to the Yorktown carrier task group. On March 14, she departed Ulithi with the fast carriers and steamed to the Japanese mainland. During strikes against targets along the coast of the Inland Sea of Japan beginning March 18, Missouri shot down four Japanese aircraft.

Raids against airfields and naval bases near the Inland Sea and southwestern Honshu continued. The Wasp was struck by a suicide plane on March 19 but resumed flight operations within an hour. Two bombs penetrated the hangar deck and decks aft of the carrier Franklin, leaving her dead in the water within 50 miles of the Japanese mainland. The cruiser Pittsburgh towed Franklin until she regained a speed of 14 knots. Missouri's carrier task group provided cover for Franklin's retreat toward Ulithi until March 22, then set course for pre-invasion strikes and bombardment of Okinawa.

Missouri joined the fast battleships of TF 58 in bombarding the southeast coast of Okinawa on March 24, an action intended to draw enemy strength from the west coast beaches, the actual site of invasion landings. Missouri rejoined the screen of the carriers as Marine and Army units stormed the shores of Okinawa on the morning of April 1. Planes from the carriers destroyed a special Japanese attacking force led by the battleship Yamato on April 7. Yamato, the world's largest battleship, was sunk, along with a cruiser and a destroyer. Three other enemy destroyers were heavily damaged and scuttled. The four remaining destroyers, sole survivors of the attacking fleet, were damaged and retreated to Sasebo.

On April 11, Missouri opened fire on a low-flying suicide plane, which penetrated the curtain of her shells and crashed just below her main deck level. The starboard wing of the plane was thrown far forward, starting a gasoline fire at 5-inch Gun Mount No. 3. However, the battleship suffered only superficial damage, and the fire was quickly brought under control.

Around 23:05 on April 17, Missouri detected an enemy submarine 12 miles from her formation. Her report initiated a hunter-killer operation by the carrier Bataan and four destroyers, which sank the Japanese submarine I-56.

Missouri was detached from the carrier task force off Okinawa on May 5 and sailed for Ulithi. During the Okinawa campaign, she shot down five enemy planes, assisted in the destruction of six others, and scored one probable kill. She helped repel 12 daylight and four night attacks on her carrier task group. Her shore bombardment destroyed several gun emplacements and many other military, governmental, and industrial structures.

Missouri arrived at Ulithi on May 9 and then proceeded to Apra Harbor, Guam, on May 18. That afternoon, Admiral William F. Halsey Jr., Commander

3rd Fleet, raised his flag on Missouri. She departed the harbor on May 21, and by May 27, was again conducting shore bombardment against Japanese positions on Okinawa. Missouri then led the mighty 3rd Fleet in strikes on airfields and installations on Kyushu on June 2

Here she prepared to lead the 3d Fleet in strikes at the heart of Japan from within its home waters. The mighty fleet set a northerly course 8 July to approach the Japanese mainland. Raids took Tokyo by surprise 10 July, followed by more devastation at the juncture of Honshu and Hokkaido 13 and 14 July. For the first time, a naval gunfire force wrought destruction On a major installation within the home islands when Missouri closed the shore to join in a bombardment 15 July that rained destruction on the Nihon Steel Co. and the wanishi Ironworks at Muroran, Hokkaido.

During the night of 17-18 July Missouri bombarded industrial targets in the Hichiti area. Honshu. Inland Sea aerial strikes continued through 25 July, and Missouri i guarded the carriers as they struck hard blows at the Japanese capital. As July ended the Japanese no longer had any home waters. Missouri had led her fleet to gain control of the air and sea approaches to the very shores of Japan.

Strikes on Hokkaido and northern Honshu resumed 9 August, the day the second atomic bomb was dropped. Next day, at 2054, Missouri men were electrified by the unofficial news that Japan was ready to surrender, provided that the Emperor's prerogatives as a sovereign

ruler were not compromised. Not until 0745, 15 August, was word received that President Truman had announced Japan's acceptance of unconditional surrender.

Adm. Sir Bruce Fraser, RN (Commander, British Pacific Fleet) boarded Missouri 16 August, and conferred the order Knight of the British Empire upon Admiral Halsey. Missouri transferred a landing party of 200 officers and men to battleship Iowa for temporary duty with the initial occupation force for Tokyo 21 August. Missouri herself entered Tokyo Bay early 29 August to prepare for the normal surrender ceremony.

High-ranking military officials of all the Allied Powers were received on board 2 September. Fleet Adm. Chester Nimitz boarded shortly after 0800, and General of the Army Douglas MacArthur (Supreme Commander for the Allies) came on board at 0.843. The Japanese representatives, headed by Foreign Minister Mamoru Shigemitsu, arrived at 0866. At 0902 General MacArthur stepped before a battery of microphones and the 23-minute surrender ceremony was broadcast to the waiting world. By 0930 the Japanese emissaries had departed.

The afternoon of 5 September Admiral Halsey transferred his flag to battleship South Dakota. Early next day Missouri departed Tokyo Bay to receive homeward bound passengers at Guam, thence sailed UN escorted for Hawaii. She arrived Pearl Harbor 20 September and flew Admiral Nimitz flag on the afternoon of 28 September for a reception.

The next day Missouri departed Pearl Harbor bound for the eastern seaboard of the United States. She reached New York City 23 October and broke the flag of Adm. Jonaq Ingrnm, commander in chief, Atlantic Fleet, .Missouri boomed out a 21-gun salute 27 October as President Truman boarded for Navy day ceremonies. In his address the President stated that "control of our sea approaches and of the skies above them is still the key to our freedom and to our ability to help enforce the peace of the world."

After overhaul in the New York Naval Shipyard and a training cruise to Cuba, Missouri returned to New York. The afternoon of 21 March 1948 she received the remains of the Turkish Ambassador to the United States, Melmet Munir Ertegun. She departed 22 March for Gibraltar and 5 April anchored in the Bosphorus off Istanbul. She rendered full honors, including the firing of a 19-gun salute during both the transfer of the remains of the late Ambassador and the funeral ashore.

Missouri departed Istanbul 9 April and entered Phaleron Bay, Piraeus, Greece, the following day for an overwhelming welcome by Greek government officials and people. She had arrived in a year when there were ominous Russian overtures and activities in the entire Balkan area. Greece had become the scene of a Communist-inspired civil war, as Russia sought every possible extension of Soviet influence throughout the Mediterranean region. Demands were made that Turkey grant the Soviets a base of seapower in the Dodecanese Islands and Joint control of the Turkish Straits leading from the Black Sea into the Mediterranean.

The voyage of Missouri to the eastern Mediterranean gave comfort to both Greece and Turkey. News media proclaimed her a symbol of U.S. interest in preserving Greek and Turkish liberty. With an August decision to deploy a strong fleet to the Mediterranean, it became obvious that the United States intended to use her naval sea and air power to stand firm against the tide of Soviet subversion.

Missouri departed Piraeus 26 April, touching at Algiers and Tangiers before arriving Norfolk 9 May. She departed for Culebra Island 12 May to join Admiral Mitscher's.8th Fleet in the Navy's first large-scale postwar Atlantic training maneuvers. The battleship returned to

New York City 27 May, and spent the next gear steaming Atlantic coastal waters north to the Davis Straits and south to the Caribbean on various Atlantic command training exercises.

Missouri arrived Rio de Janeiro 30 August 1947 for the In,ter-American Conference for the Maintenance of Hemisphere Peace and Security. President Truman boarded 2 September to celebrate the signing of the Rio Treaty which broadened the Monroe Doctrine, stipulating that an attack on one of the signatory American States would be considered an attack on all.

The Truman family boarded Missouri 7 September to return to the United States and debarked at Norfolk 19 September. Overhaul in New York (23 September to 10 March 1948) was followed by refresher training at Guantanamo Bay. Summer 1948 was devoted to midshipman and reserve training cruises. The battleship departed Norfolk 1 November for a second 3-week Arctic cold weather training cruise to the Davis Straits. The next 2 years Missouri participated in Atlantic command exercises ranging from the New England coast to the Caribbean, alternated with two midshipman summer training cruises. She was overhauled at Norfolk Naval Shipyard 23 September 1949 to 17 January 1950.

Around 23:05 on April 17, Missouri detected an enemy submarine 12 miles from her formation. Her report initiated a hunter-killer operation by the carrier Bataan and four destroyers, which sank the Japanese submarine I-56.

Missouri was detached from the carrier task force off Okinawa on May 5 and sailed for Ulithi. During the Okinawa campaign, she shot down five enemy planes, assisted in the destruction of six others, and scored one

probable kill. She helped repel 12 daylight and four night attacks on her carrier task group. Her shore bombardment destroyed several gun emplacements and many other military, governmental, and industrial structures.

Missouri arrived at Ulithi on May 9 and then proceeded to Apra Harbor, Guam, on May 18. That afternoon, Admiral William F. Halsey Jr., Commander of the 3rd Fleet, raised his flag on Missouri. She departed the harbor on May 21, and by May 27, was again conducting shore bombardment against Japanese positions on Okinawa. Missouri then led the mighty 3rd Fleet in strikes on airfields and installations on Kyushu on June 2.

Here she prepared to lead the 3rd Fleet in strikes at the heart of Japan from within its home waters. The mighty fleet set a northerly course on July 8 to approach the Japanese mainland. Raids took Tokyo by surprise on July 10, followed by more devastation at the juncture of Honshu and Hokkaido on July 13 and 14. For the first time, a naval gunfire force wrought destruction on a major installation within the home islands when Missouri closed the shore to join in a bombardment on July 15 that rained destruction on the Nihon Steel Co. and the Wanishi Ironworks at Muroran, Hokkaido.

During the night of July 17-18, Missouri bombarded industrial targets in the Hichiti area of Honshu. Inland Sea aerial strikes continued through July 25, and Missouri guarded the carriers as they struck hard blows at the Japanese capital. As July ended, the Japanese no longer had any safe home waters. Missouri had led her fleet to gain control of the air and sea approaches to the very shores of Japan.

Strikes on Hokkaido and northern Honshu resumed on August 9, the day the second atomic bomb was dropped. The next day, at 20:54, Missouri's crew was electrified by the unofficial news that Japan was ready to surrender, provided that the Emperor's prerogatives as a sovereign ruler were not compromised. Not until 07:45 on August 15 was word received that President Truman had announced Japan's acceptance of unconditional surrender.

Admiral Sir Bruce Fraser, RN (Commander, British Pacific Fleet), boarded Missouri on August 16, and conferred the order of Knight of the British Empire upon Admiral

Halsey. Missouri transferred a landing party of 200 officers and men to the battleship Iowa for temporary duty with the initial occupation force for Tokyo on August 21. Missouri herself entered Tokyo Bay early on August 29 to prepare for the formal surrender ceremony.

High-ranking military officials of all the Allied Powers were received on board on September 2. Fleet Admiral Chester Nimitz boarded shortly after 08:00, and General of the Army Douglas MacArthur (Supreme Commander for the Allies) came on board at 08:43. The Japanese representatives, headed by Foreign Minister Mamoru Shigemitsu, arrived at 08:56. At 09:02, General MacArthur stepped before a battery of microphones, and the 23-minute surrender ceremony was broadcast to the waiting world. By 09:30, the Japanese emissaries had departed.

The afternoon of September 5, Admiral Halsey transferred his flag to the battleship South Dakota. Early the next day, Missouri departed Tokyo Bay to receive homeward-bound passengers at Guam, then sailed unescorted for Hawaii. She arrived in Pearl Harbor on September 20 and flew Admiral Nimitz's flag on the afternoon of September 28 for a reception.

The next day, Missouri departed Pearl Harbor bound for the eastern seaboard of the United States. She reached New York City on October 23 and broke the flag of Admiral Jonas Ingram, Commander in Chief, Atlantic Fleet. Missouri boomed out a 21-gun salute on October 27 as President Truman boarded for Navy Day ceremonies. In his address, the President stated that 'control of our sea approaches and of the skies above them is still the key to our freedom and to our ability to help enforce the peace of the world.'

After an overhaul in the New York Naval Shipyard and a training cruise to Cuba, Missouri returned to New York. The afternoon of March 21, 1948, she received the remains of the Turkish Ambassador to the United States, Mehmet Munir Ertegun. She departed on March 22 for

Gibraltar and anchored in the Bosphorus off Istanbul on April 5. She rendered full honors, including the firing of a 19-gun salute during both the transfer of the remains of the late Ambassador and the funeral ashore.

Missouri departed Istanbul on April 9 and entered Phaleron Bay, Piraeus, Greece, the following day to an overwhelming welcome by Greek government officials and the people. She had arrived in a year when there were ominous Russian overtures and activities in the entire Balkan area. Greece had become the scene of a Communist-inspired civil war, as Russia sought every possible extension of Soviet influence throughout the Mediterranean region. Demands were made that Turkey grant the Soviets a base of seapower in the Dodecanese Islands and joint control of the Turkish Straits leading from the Black Sea into the Mediterranean.

The voyage of Missouri to the eastern Mediterranean gave comfort to both Greece and Turkey. News media proclaimed her a symbol of U.S. interest in preserving Greek and Turkish liberty. With an August decision to deploy a strong fleet to the Mediterranean, it became obvious that the United States intended to use her naval sea and air power to stand firm against the tide of Soviet subversion.

Missouri departed Piraeus on April 26, touching at Algiers and Tangiers before arriving in Norfolk on May 9. She departed for Culebra Island on May 12 to join Admiral Mitscher's 8th Fleet in the Navy's first large-scale postwar Atlantic training maneuvers. The battleship returned to New York City on May 27, and spent the next year steaming Atlantic coastal waters north to the Davis Straits and south to the Caribbean on various Atlantic command training exercises.

Missouri arrived in Rio de Janeiro on August 30, 1948, for the Inter-American Conference for the Maintenance

of Hemisphere Peace and Security. President Truman boarded on September 2 to celebrate the signing of the Rio Treaty, which broadened the Monroe Doctrine, stipulating that an attack on one of the signatory American States would be considered an attack on all.

The Truman family boarded Missouri on September 7 to return to the United States and debarked at Norfolk on September 19. An overhaul in New York (September 23 to March 10, 1948) was followed by refresher training at Guantanamo Bay. Summer 1948 was devoted to midshipman and reserve training cruises. The battleship departed Norfolk on November 1 for a second three-week Arctic cold weather training cruise to the Davis Straits. The next two years, Missouri participated in Atlantic command exercises ranging from the New England coast to the Caribbean, alternated with two midshipman summer training cruises. She was overhauled at Norfolk Naval Shipyard from September 23, 1949, to January 17, 1950.

Now the only U.S. battleship in commission, Missouri was proceeding seaward on a training mission from Hampton Roads early on January 17 when she ran aground at a point 1.6 miles from Thimble Shoals Light, near Old Point Comfort. She traversed shoal water a distance of three ship lengths from the main channel. Lifted some 7 feet above the waterline, she stuck hard and fast. With the aid of tugs, pontoons, and an incoming tide, she was refloated on February 1.

From mid-February until August 15, Missouri conducted midshipman and reserve training cruises out of Norfolk. She departed Norfolk on August 19 to support U.N. forces in their fight against Communist aggression in Korea.

Missouri joined the U.N. just west of Kyushu on September 14, becoming the flagship of Rear Admiral A. E. Smith. The first American battleship to reach Korean waters, she bombarded Samchok on September 16 in a diversionary move coordinated with the Inchon landings. In company with the cruiser Helena and two destroyers, she helped prepare the way for the 8th Army offensive.

Missouri arrived in Inchon on September 19, and on October 10 became the flagship of Rear Admiral J. M. Higgins, commander, Cruiser Division 5. She arrived in

Sasebo on October 14, where she became the flagship of Vice Admiral A. D. Struble, Commander, 7th Fleet. After screening the carrier Valley Forge along the east coast of Korea, she conducted bombardment missions from October 12 to 26 in the Chonjin and Tanchon areas, and at Wonsan. After again screening carriers eastward of Wonsan, she moved into Hungnam on December 23 to provide gunfire support around the Hungnam defense perimeter until the last U.N. troops, the U.S. 3rd Infantry Division, were evacuated by sea on Christmas Eve.

Missouri conducted additional operations with carriers and carried out systematic shore bombardments off the east coast of Korea until 19 March 1951. She arrived in Yokosuka on 24 March, and four days later, was relieved of duty in the Far East. Departing Yokosuka on 28 March, she arrived in Norfolk on 27 April and became the flagship of Rear Adm. J. L. Holloway, Jr., Commander, Cruiser Force, Atlantic Fleet. During the summer of 1951, she engaged in two midshipman training cruises to Northern Europe. Missouri entered the Norfolk Naval Shipyard on 18 October for an overhaul, which lasted until 30 January 1952.

Following winter and spring training out of Guantanamo Bay, Missouri visited New York, then set sail from Norfolk on 9 June for another midshipman cruise. She returned to Norfolk on 4 August and entered the Norfolk Naval Shipyard to prepare for a second tour in the Korean Combat Zone.

Missouri left Hampton Roads on 11 September and arrived in Yokosuka on 17 October. She raised the flag of

Vice Adm. J. J. Clark, Commander of the 7th Fleet, on 19 October. Her primary mission was to provide seagoing artillery support by bombarding enemy targets in the Chaho-Tanchon area, at Chongjin, in the Tanchon-Sonjin area, and at Chaho, Wonsan, Hamhung, and Hungnam from 25 October through 2 January 1953.

Missouri docked at Inchon on 5 January 1953 and then sailed to Sasebo, Japan. Gen. Mark Clark, Commander in Chief, U.N. Command, and Adm. Sir Guy Russell, RN, commander of the British Far East Station, visited the battleship on 23 January. In the following weeks, Missouri resumed "Cobra" patrol along the east coast of Korea, providing direct support to troops ashore. Repeated strikes against Wonsan, Tanchon, Hungnam, and Kojo destroyed major supply routes along the eastern seaboard.

Missouri's last gun-strike mission was against the Rojo area on 25 March. She sustained a serious casualty on 26 March, when her commanding officer, Capt. Warner R. Edsall, suffered a fatal heart attack while navigating her through the submarine net at Sasebo. She was relieved as the 7th Fleet flagship on 6 April by the battleship New Jersey.

Missouri departed Yokosuka on 7 April and arrived in Norfolk on 4 May, becoming the flagship for Rear Adm. E. T. Woolridge, Commander, Battleships-Cruisers, Atlantic Fleet, on 14 May. She embarked on a midshipman training cruise on 8 June, returned to Norfolk on 4 August, and underwent an overhaul in the Norfolk Naval Shipyard from 20 November to 2 April 1954.

Now the flagship of Rear Adm. R. E. Kirby, who had succeeded Admiral Woolridge, Missouri departed Norfolk on 7 June as the flagship for the midshipman training cruise to Lisbon and Cherbourg. She returned to Norfolk on 3 August and left on the 23rd for inactivation on the west coast. After visiting Long Beach and San Francisco, Missouri arrived in Seattle on 15 September. Three days later, she entered the Puget Sound Naval Shipyard and was decommissioned on 26 February 1955, joining the Bremerton group in the Pacific Reserve Fleet.

This phase of inactivity was punctuated in 1971 when the battleship was honored with a listing on the National Register of Historic Places in the State of Washington, recognizing its historical significance.

The mid-1980s marked a pivotal change in the destiny of the Missouri. As part of the ambitious 600-ship Navy plan, the battleship was reactivated and underwent extensive modernization between 1984 and 1986. This modernization included the addition of state-of-the-art cruise missile and anti-ship missile launchers, along with a comprehensive upgrade of its electronics systems. The revitalized USS Missouri was recommissioned on May 10, 1986, poised to re-enter active service.

Missouri's reentry into operational service saw it play a critical role in several key operations. During the late 1980s, specifically from September to November, Missouri was deployed in the Persian Gulf in support of Operation

Earnest Will. This operation primarily involved escorting oil tankers amidst the heightened tensions with Iran. The battleship's capability was further demonstrated in 1991 during Operations Desert Shield and Desert Storm. Notably, Missouri had the distinction of being the first battleship to launch Tomahawk missiles in combat. Additionally, in December 1991, it participated in Operation Remembrance in Hawaii, marking the 50th anniversary of the Pearl Harbor attack, a poignant moment that tied back to its historical significance.

However, this period of renewed activity was not to last. On March 31, 1992, Missouri was decommissioned for the second and final time. Subsequently, it was removed from the Navy's ship registry. The next chapter in Missouri's storied history began in 1996 when the USS Missouri Memorial Association was selected to receive the decommissioned battleship. By June 1998, Missouri made its way to Waikiki, and by January 29, 1999, a new phase of its life commenced as it opened to the public as the Battleship Missouri Memorial. Located on Battleship Row, Ford Island, Pearl Harbor, the Missouri now serves as a living museum, offering a tangible connection to the past and a solemn reminder of the ship's historical importance and the broader narrative of naval warfare and diplomacy.

Missouri received three battle stars for her service in World War II and five for her service in Korea.

BB-64 • USS WISCONSIN

(BB-64: dp. 45,000; l. 887'3"; b. 108'3"; dr. 28'11" (mean); s. 33 k.; cpl. 1,921; a. 9 16", 20 5", 80 40mm., 49 20mm.; cl. Iowa)

The second Wisconsin (BB-64) was laid down on 25 January 1941 at the Philadelphia Navy Yard; launched on 7 December 1943, sponsored by Mrs. Walter S. Goodland, and commissioned on 16 April 1944, Capt. Earl E. Stone in command.

After her trials and initial training in the Chesapeake Bay, Wisconsin departed Norfolk, VA., on 7 July 1944, bound for the British West Indies. Following her shakedown, conducted out of Trinidad, the third of the Iowa-class battleships to join the Fleet returned to her builder's yard for post-shakedown repairs and alterations.

On 24 September 1944, Wisconsin sailed for the west coast, transited the Panama Canal, and reported for duty with the Pacific Fleet on 2 October. The battleship later moved to Hawaiian waters for training exercises and then headed for the Western Carolines. Upon reaching Ulithi on 9 December, she joined Admiral William F. Halsey's 3rd Fleet.

The powerful new warship had arrived at a time when the reconquest of the Philippines was well underway. As a part of that movement, the planners had envisioned landings on the southwest coast of Mindoro south of Luzon. From that point, American forces could threaten Japanese shipping lanes through the South China Sea.

The day before the amphibians assaulted Mindoro, the 3rd Fleet's Fast Carrier Task Force (TF) 38—supported in part by Wisconsin—rendered Japanese facilities at Manila largely useless. Between 14 and 16 December, TF 38's naval aviators secured complete tactical surprise and quickly won complete mastery of the air and sank or destroyed 27 Japanese vessels, damaged 60 more; destroyed 269 planes; and bombed miscellaneous ground installations.

The next day, however, the weather soon turned sour for Halsey's sailors. A furious typhoon struck his fleet, catching many ships refueling and with little ballast in their nearly dry bunkers. Three destroyers— Hull (DD-350), Monaghan (DD-354), and Spence (DD-512)—capsized and sank. Wisconsin proved her seaworthiness as she escaped the storm unscathed.

As heavily contested as they were, the Mindoro operations proved only the introduction to another series of calculated blows aimed at the occupying Japanese in the Philippines. For Wisconsin, her next operation was the occupation of Luzon. By-passing the southern beaches, American amphibians went ashore at Lingayen Gulf—the scene of the Japanese landings nearly three years before.

Wisconsin—armed with heavy antiaircraft batteries—performed escort duty for TF 38's fast carriers during air strikes against Formosa, Luzon, and the Nansei Shoto, to neutralize Japanese forces there and to cover the unfolding Lingayen Gulf operations. Those strikes, lasting from 3 to 22 January 1945, included a thrust into the South China Sea, in the hope that major units of the Japanese Navy could be drawn into battle.

Air strikes between Saigon and Camranh Bay, Indochina, on 12 January resulted in severe losses for the enemy. TF 38's warplanes sank 41 ships and damaged 31 in two convoys they encountered. In addition, they heavily damaged docks, storage areas, and aircraft facilities. At least 112 enemy planes would never again see operational service. Formosa, already struck on 3 and 4 January, again fell victim to the marauding American airmen, being smashed again on 9, 15, and 21 January. Soon, Hong Kong, Canton, and Hainan Island felt the brunt of TF 38's power. Besides damaging and sinking Japanese shipping, American planes from the task force set the Canton oil refineries afire and blasted the Hong Kong Naval Station. They also raided Okinawa on 22 January, considerably lessening enemy air activities that could threaten the Luzon landings.

Subsequently assigned to the 5th Fleet—when Admiral Spruance relieved Admiral Halsey as Commander of the Fleet—Wisconsin moved northward with the redesignated TF 58 as the carriers headed for the Tokyo area. On 16 February 1945, the task force approached the Japanese coast under cover of adverse weather conditions and achieved complete tactical surprise. As a result, they shot down 322

enemy planes and destroyed 177 more on the ground. Japanese shipping—both naval and merchant—suffered drastically, too, as did hangars and aircraft installations. Moreover, all this damage to the enemy had cost the American Navy only 49 planes.

The task force moved to Iwo Jima on 17 February to provide direct support for the landings slated to take place on that island on the 19th. It revisited Tokyo on the 25th and, the next day, hit the island of Hachino off the coast of Honshu. During these raids, besides causing heavy damage to ground facilities, the American planes sent five small vessels to the bottom and destroyed 158 planes.

On 1 March, reconnaissance planes flew over the island of Okinawa, taking last-minute intelligence photographs to be used in planning the assault on that island. The next day, cruisers from TF 58 shelled Okino Daito Shima in training for the forthcoming operation. The force then retired to Ulithi for replenishment.

Wisconsin's task force stood out of Ulithi on 14 March, bound for Japan. The mission of that group was to eliminate airborne resistance from the Japanese homeland to American forces off Okinawa. Enemy fleet units at Kure and Kobe, on southern Honshu, reeled under the impact of the explosive blows delivered by TF 58's airmen. On 18 and 19 March, from a point 100 miles southwest of Kyushu, TF 58 hit enemy airfields on that island. However, the Japanese drew blood during that action when kamikazes crashed into Franklin (CV-17) on the 19th and seriously damaged that fleet carrier.

That afternoon, the task force retired from Kyushu, screening the blazing and battered flattop. In doing so, the screen downed 48 attackers. At the conclusion of the operation, the force felt that it had achieved its mission of prohibiting any large-scale resistance from the air to the slated landings on Okinawa.

On the 24th, Wisconsin trained her 16-inch rifles on targets ashore on Okinawa. Together with the other battlewagons of the task force, she pounded Japanese positions and installations in preparation for the landings. Although fierce, Japanese resistance was doomed to fail by dwindling numbers of aircraft and trained pilots to man them. In addition, the Japanese fleet, steadily hammered by air attacks from 5th Fleet aircraft, found itself confronted by a growing, powerful, and determined enemy. On 17 April, the undaunted enemy battleship Yamato, with her 18.1-inch guns, sortied to attack the American invasion fleet off Okinawa. Met head-on by a swarm of carrier planes, Yamato, the light cruiser Yahagi, and four destroyers went to the bottom, the victims of massed air power. Never again would the Japanese fleet present a major challenge to the American fleet in the war in the Pacific.

While TF 58's planes were off dispatching Yamato and her consorts to the bottom of the South China Sea, enemy aircraft struck back at American surface units. Combat air patrols (CAP) knocked down 16 enemy planes, and ships' gunfire accounted for another three, but not before one kamikaze penetrated the CAP and screen to crash on the flight deck of the fleet carrier Hancock (CV-19). On 11 April, the "Divine Wind" renewed its efforts; and only drastic maneuvers and heavy barrages of gunfire saved the task force. None of the fanatical pilots achieved any direct hits, although near misses, close aboard, managed to cause

some minor damage. Combat air patrols bagged 17 planes, and ships' gunfire accounted for an even dozen. The next day, 151 enemy aircraft committed harakiri into TF 58, but Wisconsin, bristling with 5-inch, 40-millimeter, and 20-millimeter guns, together with other units of the screens for the vital carriers, kept the enemy at bay or destroyed him before he could reach his targets.

Over the days that ensued, American task force planes hit Japanese facilities and installations in the enemy's homeland. Kamikazes, redoubling their efforts, managed to crash into three carriers on successive days—Intrepid (CV-11), Bunker Hill (CV-17), and Enterprise (CV-6).

By 4 June, a typhoon was swirling through the Fleet. Wisconsin rode out the storm unscathed, but three cruisers, two carriers, and a destroyer suffered serious damage. Offensive operations were resumed on 8 June with a final aerial assault on Kyushu. Japanese aerial response was pitifully small, 29 planes were located and destroyed. On that day, one of Wisconsin's floatplanes landed and rescued a downed pilot from the carrier Shangri-La (CV-38).

Wisconsin ultimately put into Leyte Gulf and dropped anchor there on 13 June for repairs and replenishment. Three weeks later, on 1 July, the battleship and her consorts sailed once more for Japanese home waters for carrier air strikes on the enemy's heartland. Nine days later, carrier planes from TF 38 destroyed 72 enemy aircraft on the ground and smashed industrial sites in the Tokyo area. So little was the threat from the dwindling Japanese air arm that the Americans made no attempt whatever to conceal the location of their armada which was operating off her shores with impunity.

On the 15th, Wisconsin again unlimbered her main battery, hurling 16-inch shells shoreward at the steel mills and oil refineries at Muroran, Hokkaido. Two days later, she wrecked industrial facilities in the Hitachi Mito area, on the coast of Honshu, northeast of Tokyo itself. During that bombardment, British battleships of the Eastern Fleet contributed their heavy shellfire. By that point in the war, Allied warships were able to shell the Japanese homeland almost at will.

Task Force 38's planes subsequently blasted the Japanese naval base at Yokosuka, and put one of the two remaining

Japanese battleships—the former fleet flagship Nagato—out of action. On 24 and 25 July, American carrier planes visited the Inland Sea region, blasting enemy sites on Honshu, Kyushu, and Shikoku. Kure then again came under attack. Six major fleet units were located there and badly damaged, marking the virtual end of Japanese sea power.

Over the weeks that ensued, TF 38 continued its raids on Japanese industrial facilities, airfields, and merchant and naval shipping. Admiral Halsey's airmen visited destruction upon the Japanese capital for the last time on 13 August 1945. Two days later, the Japanese capitulated. World War II was over at last.

Wisconsin, as part of the occupying force, arrived at Tokyo Bay on 6 September, three days after the formal surrender occurred on board the battleship Missouri (BB-63). During Wisconsin's brief career in World War II, she had steamed 105,831 miles since commissioning, had shot down three enemy planes, had claimed assists on four occasions, and had fueled her screening destroyers on some 250 occasions.

Shifting subsequently to Okinawa, the battleship embarked homeward-bound GIs on 22 September, as part of the "Magic Carpet" operation staged to bring soldiers, sailors, and marines home from the far-flung battlefronts of the Pacific. Departing Okinawa on 23 September, Wisconsin reached Pearl Harbor on 4 October, remaining there for five days before she pushed on for the west coast on the last leg of her stateside-bound voyage. She reached San Francisco on 16 October.

Heading for the east coast of the United States soon after the start of the new year, 1946, Wisconsin transited the Panama Canal between 11 and 13 January, and reached Hampton Roads, VA., on the 18th. Following a cruise south to Guantanamo Bay, Cuba, the battleship entered the Norfolk Naval Shipyard for overhaul. After repairs and alterations that consumed the summer months, Wisconsin sailed for South American waters.

Over the weeks that ensued, the battleship visited Valparaiso, Chile, from 1 to 6 November; Callao, Peru, from 9 to 13 November; Balboa, Canal Zone, from 16 to 20 November; and La Guaira, Venezuela, from 22 to 26 November, before returning to Norfolk on 2 December 1946.

Wisconsin spent nearly all of 1947 as a training ship, taking naval reservists on two-week cruises throughout the year. Those voyages commenced at Bayonne, N.J., and saw visits conducted at Guantanamo Bay, Cuba, and the Panama Canal Zone. While underway at sea, the ship would perform various drills and exercises before the cruise would end where it had started, at Bayonne. During June and July of 1947, Wisconsin took Naval Academy midshipmen on cruises to northern European waters.

In January 1948, Wisconsin joined the Atlantic Reserve Fleet at Norfolk, for inactivation. Placed out of commission, in reserve, on 1 July 1948, Wisconsin was assigned to the Norfolk group of the Atlantic Reserve Fleet.

Her sojourn in "mothballs," however, was comparatively brief because of the North Korean invasion of South Korea in late June 1950. Wisconsin was recommissioned on 3 March 1951, Capt. Thomas Burrowes in command. After shakedown training, the revitalized battleship conducted two midshipmen training cruises taking the officers-to-be to Edinburgh, Scotland; Lisbon, Portugal; Halifax, Nova Scotia; New York City, and Guantanamo Bay, Cuba, before she returned to Norfolk.

Wisconsin departed Norfolk on 25 October 1951 bound for the Pacific. She transited the Panama Canal on the 29th and reached Yokosuka, Japan, on 21 November. There, she relieved New Jersey (BB-62) as flagship for Vice Admiral H. M. Martin, Commander, 7th Fleet.

On the 26th, with Vice Admiral Martin and Rear Admiral F. P. Denebrink, Commander, Service Force Pacific, embarked, Wisconsin departed Yokosuka for Korean waters to support the fast carrier operations of TF 77. She left the company of the carrier force on 2 December and, screened by the destroyer Wiltsie (DD-716), provided gunfire support for the Republic of Korea (ROK) Corps in the Kansong-Kosong area. After disembarking Admiral Denebrink on 3 December at Kangnung, the battleship resumed station on the Korean "bombline," providing gunfire support for the American 1st Marine Division. Wisconsin's shellings accounted for a tank, two gun emplacements, and a building. She continued her gunfire support task for the 1st Marine Division and 1st ROK Corps through 6 December, accounting for enemy bunkers, artillery positions, and troop concentrations. On one occasion during that time, the battleship received a request for call-fire support and provided three star shells for the 1st ROK Corps, illuminating a communist attack that was consequently repulsed with considerable enemy casualties.

After being relieved on the gunline by the heavy cruiser St. Paul (CA-73) on 6 December, Wisconsin retired only briefly from gunfire support duties. She resumed them, however, in the Kansong-Kosong area on 11 December, screened by the destroyer Twining (DD-540). The following day, 12 December, saw the embarkation in Wisconsin of Rear Admiral H. R. Thurber, Commander, Battleship Division 2. The admiral came on board via helicopter, incident to his inspection trip in the Far East.

The battleship continued naval gunfire support duties on the "bombline," shelling enemy bunkers, command posts, artillery positions, and trench systems through 14 December. She departed the "bombline" on that day to render special gunfire support duties in the Kojo area, blasting coastal targets in support of United Nations (UN) troops ashore. That same day, she returned to the Kansong-Kosong area. On the 15th, she disembarked Admiral Thurber by helicopter. The next day, Wisconsin departed Korean waters, heading for Sasebo to rearm.

Returning to the combat zone on the 17th, Wisconsin embarked United States Senator Homer Ferguson of Michigan on the 18th. That day, the battleship supported the 11th ROK division with night illumination fire that enabled the ROK troops to repulse a communist assault with heavy enemy casualties. Departing the "bombline" on the 19th, the battleship later that day transferred her distinguished passenger, Senator Ferguson, by helicopter to the carrier Valley Forge (CV-45).

Wisconsin next participated in a coordinated air-surface bombardment of Wonsan to neutralize preselected targets. She shifted her bombardment station to the western end of Wonsan harbor, hitting boats and small craft in the inner swept channel during the afternoon. Such activities helped to forestall any communist attempts to assault the friendly held islands in the Wonsan area. Wisconsin then made an anti-boat sweep to the north, utilizing her 5-inch batteries on suspected boat concentrations. She then provided gunfire support to UN troops operating at the "bombline" until three days before Christmas 1951, when she rejoined the carrier task force.

On 28 December, Francis Cardinal Spellman—on a Korean tour over the Christmas holidays—visited the ship, coming on board by helicopter to celebrate Mass for the Catholic members of the crew. The distinguished prelate departed the ship by helicopter off Pohang. Three days later, on the last day of the year, Wisconsin put into Yokosuka.

Wisconsin departed that Japanese port on 8 January 1952 and headed for Korean waters once more. She reached Pusan the following day and entertained the President of South Korea, Syngman Rhee, and his wife on the 10th. President and Mrs. Rhee received full military honors as they came on board, and he reciprocated by awarding Vice Admiral Martin the ROK Order of Military Merit.

Wisconsin returned to the "bombline" on January 11 and, over the ensuing days, delivered heavy gunfire support for the 1st Marine Division and the 1st ROK

Corps. As before, her primary targets were command posts, shelters, bunkers, troop concentrations, and mortar positions. As before, she stood ready to deliver call-fire support as needed. One such occasion occurred on January 14 when she shelled enemy troops in the open at the request of the ROK 1st Corps.

Rearming at Sasebo and once more joining TF 77 off the coast of Korea soon thereafter, Wisconsin resumed support at the "bombline" on January 23. Three days later, she shifted once more to the Kojo region to participate in a coordinated air and gun strike. That same day, the battleship returned to the "bombline" and shelled the command post and communications center for the 15th North Korean Division during call-fire missions for the 1st Marine Division.

Returning to Wonsan at the end of January, Wisconsin bombarded enemy guns at Hodo Pando before she was rearmed at Sasebo. The battleship rejoined TF 77 on February 2 and, the next day, blasted railway buildings and marshaling yards at Hodo Pando and Kojo before rejoining TF 77. After replenishment at Yokosuka a few days later, she returned to the Kosong area and resumed gunfire support. During that time, she destroyed railway bridges and a small shipyard, besides conducting call fire missions on enemy command posts, bunkers, and personnel shelters, making numerous cuts on enemy trench lines in the process.

On February 25, Wisconsin arrived at Pusan, where Vice Admiral Shon, the ROK Chief of Naval Operations; United States Ambassador J. J. Muccio; and Rear Admiral Scott-Montcrief, Royal Navy, Commander, Task Group 95.12, visited the battleship. Departing that South Korean port the following day, Wisconsin reached Yokosuka on March 2. A week later, she shifted to Sasebo to prepare to return to Korean waters.

Wisconsin arrived in Songjin, Korea, on March 15, 1952, and concentrated her gunfire on enemy railway transport. Early that morning, she destroyed a communist troop train trapped outside of a destroyed tunnel. That afternoon, she received the first direct hit in her history when one of four shells from a communist 155-millimeter gun battery struck the shield of a starboard 40-millimeter

mount. Although little material damage resulted, three men were injured. Almost as if the victim of a personal affront, Wisconsin subsequently blasted that battery to oblivion with a 16-inch salvo before continuing her mission. After lending a hand to support the 1st Marine Division once more with her heavy rifles, the battleship returned to Japan on March 19.

Relieved as flagship of the 7th Fleet on April 1 by sistership Iowa (BB-61), Wisconsin departed Yokosuka, bound for the United States. En route home, she touched briefly at Guam, where she took part in the successful test of the Navy's largest floating drydock on April 4 and 5, marking the first time that an Iowa class battleship had ever utilized that type of facility. She continued her homeward-bound voyage via Pearl Harbor and arrived at Long Beach, Calif., on April 19. She then sailed for the east coast, her destination. Norfolk.

Early in June 1952, Wisconsin resumed her role as a training ship, taking midshipmen to Greenock, Scotland; Brest, France; and Guantanamo Bay, Cuba, before returning to Norfolk. She departed Hampton Roads on August 25 and participated in a North Atlantic Treaty Organization (NATO) exercise, Operation "Mainbrace," which commenced at Greenock and extended as far north as Oslo, Norway. After her return to Norfolk, Wisconsin underwent an overhaul in the naval shipyard there. She then engaged in local training evolutions until February 11, 1953, when she sailed for Cuban waters for refresher training. She visited Newport, R.I., and New York City before returning to Norfolk late in April.

Following another midshipman's training cruise to Rio de Janeiro, Brazil; Port-of-Spain, Trinidad; and Guantanamo Bay, Wisconsin, they were put into the Norfolk Naval Shipyard on August 4 for a brief overhaul. A little over a month later, upon the conclusion of that period of repairs and alterations, the battleship departed Norfolk on September 9, bound for the Far East.

Sailing via the Panama Canal to Japan, Wisconsin relieved New Jersey (BB-62) as the 7th Fleet flagship on October 12. During the months that followed, Wisconsin visited the Japanese ports of Kobe, Sasebo, Yokosuka Otaru, and Nagasaki. She spent Christmas in Hong Kong and was ultimately relieved of flagship duties on April 1, 1954, and returned to the United States soon thereafter, reaching Norfolk via Long Beach and the Panama Canal on May 4, 1954.

Entering the Norfolk Naval Shipyard on June 11, Wisconsin underwent a brief overhaul and commenced a midshipman training cruise on July 12. After revisiting Greenock, Brest, and Guantanamo Bay, the ship returned to the Norfolk Naval Shipyard for repairs. Shortly thereafter, Wisconsin participated in Atlantic Fleet exercises as flagship for Commander, 2d Fleet. Departing Norfolk in January 1955, Wisconsin took part in Operation "Springboard," during which time she visited Port-au-Prince, Haiti. Then, upon returning to Norfolk, the battleship conducted another midshipman's cruise that summer, visiting Edinburgh, Copenhagen, Denmark, and Guantanamo Bay before returning to the United States.

Upon completion of a major overhaul at the New York Naval Shipyard, Wisconsin headed south for refresher training in the Caribbean, later taking part in another "Springboard" exercise. During that cruise, she again visited Port-au-Prince and added Tampico, Mexico, and Cartagena, Colombia, to her list of ports of call. She returned to Norfolk on the last day of March, 19566, for local operations.

Throughout April and into May, Wisconsin operated locally off the Virginia capes. On May 6, the battleship collided with the destroyer Eaton (DDE-510) in a heavy fog. Wisconsin put into Norfolk with extensive damage to her bow and, one week later, entered drydock at the Norfolk Naval Shipyard. A novel expedient speeded her repairs and enabled the ship to carry out her scheduled midshipman training cruise that summer. A 120-ton, 68-foot long section of the bow of the uncompleted battleship Kentucky was transported by barge in one section from the Newport News Shipbuilding and Drydock Corp., Newport News, VA., across Hampton Roads to the Norfolk Naval Shipyard. Working round-the-clock, Wisconsin's ship's force and

shipyard personnel completed the operation, which grafted the new bow on the old battleship in a mere 16 days. On June 28, 1956, the ship was ready for sea.

Embarking 700 NROTC midshipmen, representing 62 colleges and universities throughout the United States, Wisconsin departed Norfolk on July 9, bound for Spain. Reaching Barcelona on the 20th, the battleship next called at Greenock and Guantanamo Bay before returning to Norfolk on the last day of August. That autumn, Wisconsin participated in Atlantic Fleet exercises off the coast of the Carolinas, returning to port on November 8, 1956. Entering the Norfolk Naval Shipyard a week later, the battleship underwent major repairs that were not finished until January 2, 1957.

After local operations off the Virginia capes from 3 to January 4 and from the 9th to the 11th, W18consin departed Norfolk on the 15th, reporting to Commander Fleet Training Group at Guantanamo Bay. Breaking the two-starred flag of Rear Admiral Henry Crommelin, Commander, Battleship Division 2, Wisconsin, served as Admiral Crommelin's flagship during the ensuing shore bombardment practices and other exercises held off the isle of Culebra, Puerto Rico, from 2 to February 4, 1967. Sailing for Norfolk upon completion of the training period, the battleship arrived on February 7.

The warship conducted a brief period of local operations off Norfolk before she sailed for the Mediterranean on March 27. Reaching Gibraltar on April 5, she pushed on that day to rendezvous with TF 60 in the Aegean Sea. She then proceeded with that force to Xeros Bay, Turkey, arriving there on April 11 for NATO Exercise "Red Pivot."

Departing Xeros Bay on April 14, she arrived at Naples four days later. After a week's visit—during which she was visited by Italian dignitaries—Wisconsin conducted exercises in the eastern Mediterranean. In the course of those operational training evolutions, she rescued a pilot and crewman who survived the crash of a plane from the carrier Forre8tal (CVA-59). Two days later, Vice Admiral

Charles R. Brown, Commander of the 6th Fleet, came on board for an official visit by Highline and departed via the same method that day. Wisconsin reached Valencia, Spain, on May 10 and, three days later, entertained prominent civilian and military officials of the city.

Departing Valencia on the 17th, Wisconsin reached Norfolk on May 27. On that day, Rear Admiral L. S. Parks relieved Rear Admiral Crommelin as Commander Battleship Division 2. Departing Norfolk on June 19, the battleship, over the ensuing weeks, conducted a midshipman training cruise through the Panama Canal to South American waters. She transited the canal on June 26, crossed the equator on the following day, and reached Valparaiso, Chile, on July 3. Eight days later, the battleship headed back to the Panama Canal and the Atlantic.

After exercises at Guantanamo Bay and off Culebra, Wisconsin reached Norfolk on August 5 and conducted local operations that lasted into September. She then participated in NATO exercises, which took her across the North Atlantic to the British Isles. She arrived in the Clyde on September 14 and subsequently visited Brest, France, before returning to Norfolk on October 22.

Wisconsin's days as an active fleet unit were numbered, and she prepared to make her last cruise. On November 4, 1957, she departed Norfolk with a large group of prominent guests on board. Reaching New York City on November 6, the battleship disembarked her guests and, on the 8th, headed for Bayonne, N.J., to commence pre-inactivation overhaul.

Placed out of commission at Bayonne on March 8, 1958, Wisconsin joined the "Mothball Fleet" there, leaving the United States Navy without an active battleship for the first time since 1895. Subsequently taken to the Philadelphia Naval Shipyard, Wisconsin, and remained there with her sistership Iowa into 1981.

Wisconsin was recommissioned on October 22, 1988, armed with Tomahawk missiles.

Wisconsin served in Operation Desert Storm from January 15 to February 27, 1991. This marked the last time that a United States battleship ever actively participated in a foreign war.

With the collapse of the Soviet Union and the absence of a perceived threat to the United States came drastic cuts in the defense budget, and the high cost of maintaining and operating battleships as part of the United States Navy became uneconomical. As a result, the Wisconsin was decommissioned on September 30, 1991, and was stricken from the Naval Vessel Register on January 12, 1995. On October 15, 1996, she was moved to the Norfolk Naval Shipyard. On February 12, 1998, she was restored to the Naval Vessel Register. She remains berthed adjacent to Nauticus in Norfolk, Virginia.

Wisconsin earned five battle stars for her World War II service and one for the Korean War. The ship also received the Navy Unit Commendation for service during the first Gulf War.

The City of Norfolk has assumed stewardship of the Battleship Wisconsin.

Printed in Great Britain
by Amazon

50150568R00106